BEGINNING SCIENCE

BIOLOGY

B S BECKETT

Oxford University Press

D0258672

Acknowledgements

The publisher would like to thank the following for permission to reproduce photographs:

All-Sport Photographic, p. 6 (left); Ardea, p. 21 (all 3 top, top lower right, bottom left, bottom upper right), p. 23 (centre left), p. 37, p. 39, p. 94, p. 140 (bottom); Barnaby's Picture Library, p. 125; Biofotos, p. 8, p. 13 (top left, top centre, centre left, centre right, all 3 bottom), p. 15 (top right, centre right, all 3 bottom), p. 17 (top left, centre left, all 3 bottom), p. 19 (top centre, centre left, centre right, bottom left and right), p. 22 (bottom), p. 23 (all 3 top, centre, bottom left, bottom right), p. 27, p. 88, p. 92, p. 96, p. 100, p. 118, p. 121 (bottom); Dr B. Bracegirdle, p. 66 (top), p. 113; Camera Press, p. 48, p. 64, p. 150 (upper top, upper and lower bottom); Camera Talks, p. 131 (bottom right); Centre for Overseas Pest Research, p. 62 (centre); Bruce Coleman, p. 1, p. 13 (centre), p. 17 (top right), p. 54, p. 87, p. 115, p. 124, p. 141; the Daily Telegraph Colour Library, p. 151 (left); Henry Grant, p. 132; Philip Harris Biological Limited, p. 117 (bottom); Health Education Council, p. 147; Eric and David Hosking, p. 13 (top right), p. 15 (top left, top centre, centre), p. 17 (centre, centre right), p. 15 (top left, top centre, centre), p. 17 (centre, centre right), p. 19 (top right, centre), p. 21 (centre), p. 66 (bottom); Keystone Press Agency, p. 6 (right);

J. H. Kugler (Philip Harris Biological Limited), p. 111; La Leche League of Hull, p. 51; Frank W. Lane, p. 15 (centre left), p. 21 (bottom lower right), p. 22 (top), p. 23 (centre right, bottom centre), p. 28, p. 140 (top); Life © Time Inc. 1980/Colorific!, p. 137 (bottom); London Scientific Fotos, p. 68, p. 127, p. 128; N.A.S.A./ Space Frontiers, p. 42; National Association of British and Irish Millers Limited, p. 53 (top); Natural History Photographic Agency, p. 17 (top centre), p. 19 (top left, bottom centre), p. 62 (bottom); Oxford Scientific Films, p. 36, p. 45, p. 56, p. 99, p. 120, p. 122, p. 134, p. 144; Popperfoto, p. 137 (top), p. 150 (lower top); Radcliffe Infirmary, p. 71 (left), p. 147 (bottom right); Mervyn Rees, p. 148; Shell Photographic Library, p. 62 (top); St. Bartholomew's Hospital, p. 151 (right); The Sunday Times/Ian Yeomans, p. 83 (top left); Jeffrey Tabberner, p. 53 (bottom); John Topham Picture Library, p. 102 (top), p. 121 (top); Vision International, p. 83 (top right); M. I. Walker (Philip Harris Biological Limited), p. 117 (top); War on Want, p. 50; John Watney Photo Library, p. 4, p. 73, p. 74, p. 77, p. 83 (bottom), p. 85, p. 89, p. 90, p. 98, p. 131 (top centre, bottom left); C. James Webb, p. 60, p. 65, p. 86, p. 102 (bottom), p. 106, p. 114, p. 129, p. 152; Martin White, p. 71 (right).

Cover photographs: Bruce Coleman and Oxford Scientific Films.

Contents

Topic 8   Reproduction and heredity

Topic 9   Health and hygiene

Forms of life

These protists are a form of life that is neither animal nor plant

1.1   Studying living things

Biology is the study of living things, but what is meant by living? This question can be answered by making a list of the features which are shared by all living things.

Features of living things

When listing the features shared by living things it is no good including walking, talking, or even breathing, because many living things cannot walk or talk and plants do not breathe in and out. In fact there are only seven features which *all* living things have in common.

1 Movement

All living things are capable of movement. Most animals move about using legs, wings, or fins. Plants move by growing. Their roots grow down into the soil, and their shoots grow up into the air or towards a source of light.

2 Sensitivity

All living things are sensitive to certain changes in their surroundings. This means that, to a certain extent, they are aware of what is happening around them. Many animals use sense organs such as eyes and ears to find out things about their surroundings. Plants do not have sense organs but are still able to detect and respond to things like gravity, light, and water.

3 Feeding

All living things need food to provide energy and for growth. Animals get their food by eating other living or once-living things. *Herbivores* (e.g. rabbits) eat plants, *carnivores* (e.g. lions) eat other animals, and *omnivores* (e.g. most humans) eat animals and plants. Plants make their own food. They use the energy of sunlight to combine water and carbon dioxide gas to make sugar. This is called *photosynthesis*.

4 Respiration

All living things need energy: for movement, to work the organs of the body, and for growth. This energy is obtained from food by a process called *respiration*. During respiration, a number of chemical changes release energy from food, usually by combining the food with oxygen.

5 Excretion

All living things produce wastes, such as carbon dioxide (which is a waste product of respiration), water, urine, and other chemicals. *Excretion* is the name for processes which remove these wastes from the body.

6 Reproduction

All living things reproduce, to replace organisms lost by death. If a group of organisms does not reproduce fast enough to replace those which die, the group becomes *extinct*. Some very small creatures can reproduce by splitting in two. This is an example of *asexual reproduction*. Most animals are male or female and these can mate and produce young. Mating is an example of *sexual reproduction*. Plant flowers contain sexual organs which produce seeds, and the seeds grow into new plants.

7 Growth

All living things grow. Animals grow until they reach a certain adult size, but most plants can grow continuously throughout their lives.

Exercises

1   What is a biologist?

2   What do the following words mean: sensitivity, photosynthesis, respiration, excretion?

3   What are animals sensitive to, and what are plants sensitive to?

4   What is food needed for, and what use do living things make of the energy released from food by respiration?

5   What substances do living things excrete?

6   What is the difference between the way plants and animals grow?

7   A motor car moves from place to place, obtains energy by combining petrol with oxygen, and produces waste gases. Does this mean cars are alive? Look through the seven features of living things and list those which do not apply to cars.

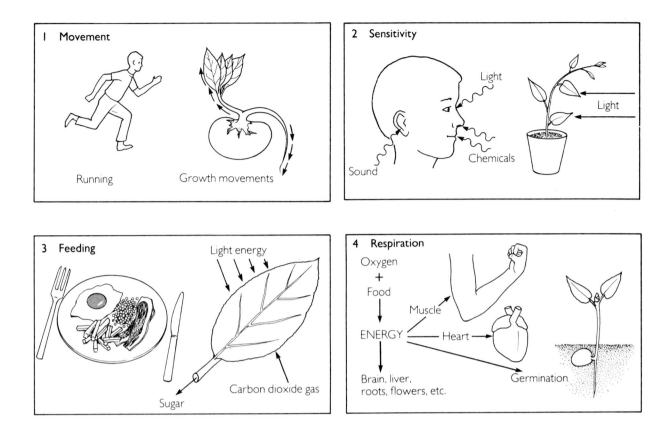

1   Movement

Running   Growth movements

2   Sensitivity

Light

Light

Sound

Chemicals

3   Feeding

Light energy

Sugar

Carbon dioxide gas

4   Respiration

Oxygen
+
Food

↓

ENERGY — Heart →

Muscle

Germination

↓

Brain, liver,
roots, flowers, etc.

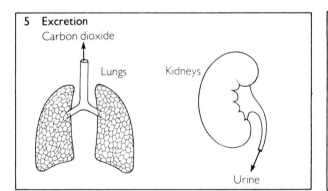

5   Excretion

Carbon dioxide

Lungs

Kidneys

Urine

6   Reproduction

Seeds

Babies

Chicks

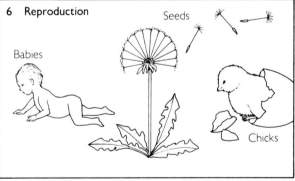

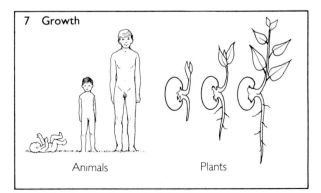

7   Growth

Animals   Plants

Fig. 1   The seven features of living things

1.2　Cells, tissues, and organs

A white blood cell

Our bodies are not continuous material, but are made up of millions of separate blobs of living matter called cells. This vast army of cells work together, performing all the tasks which keep us alive.

What is a cell?

A cell is a unit of living matter. The largest cells are bird eggs – an ostrich egg is up to 17 cm long. Bird eggs are big because they are filled with food (yolk and albumen). Most cells are microscopic in size. If a man increased in size a million times his cells would still only be the size of a cricket ball.

Unicellular organisms are those which consist of only one cell. *Amoeba* and *Paramecium* are examples illustrated in Unit 1.5.
Multicellular organisms, which include all animals and plants, consist of many cells (Figs. 3 and 4).

Parts of a cell

All cells consist of living material called **protoplasm**, and are made up of three parts (Figs. 1 and 2). They have an outer skin called the **cell membrane** which encloses a jelly-like substance called **cytoplasm**. A rounded object called the **nucleus** floats in the cytoplasm.

Cell membrane This controls the movement of substances in and out of a cell. It allows waste substances to leave the cell, and lets other chemicals, such as oxygen and food, enter the cell.

Cytoplasm Cytoplasm is 90 per cent water, plus proteins, oils, glucose sugar, vitamins, and minerals. Narrow passageways extend throughout the cytoplasm and it often contains larger fluid-filled spaces called **vacuoles**. Cytoplasm also contains grains of stored food and oil droplets.

Nucleus The nucleus is a cell's control centre. First and foremost it controls the formation and development of a cell. All multicellular organisms begin life as one cell. The nucleus and cytoplasm of this first cell divide, producing two cells, which divide producing four, and so on until there are millions of cells. After each cell is formed its nucleus controls its development: whether it becomes a blood cell, bone cell, leaf cell, etc. The nucleus also controls the chemicals which the cell manufactures.

Cells, tissues, and organs

The cells of an organism are not all alike. Most are specialized to carry out one special job. Groups of specialized cells are called **tissues**. Muscle tissue consists of cells specialized to contract and move the body. Nervous tissue consists of cells which allow nerve impulses to travel along them. An **organ** consists of several different tissues which work together. The heart is an organ consisting of muscle and nervous tissues, bound together with fibres called connective tissue.

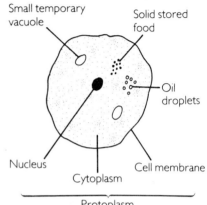

Fig. 1 Diagram of an animal cell Animal cells have small temporary vacuoles and have a greater variety of shape and function than plant cells. They never have chloroplasts or a cellulose wall.

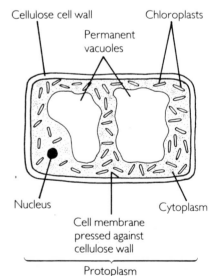

Fig. 2 Diagram of a plant cell Plant cells have a cellulose wall outside the cell membrane, large permanent vacuoles, and chloroplasts which contain chlorophyll and carry out photosynthesis.

Exercises

1 How many cells are there in a unicellular organism? Name two examples of such an organism.

2 What three things make up the protoplasm of a cell? Which part is the cell's control centre?

3 What is the function of the cell membrane?

4 Describe the composition of a cell's cytoplasm.

5 Name three different types of specialized cells. Which part of a cell determines how it will specialize?

6 Explain the difference between 'tissue' and 'organ'.

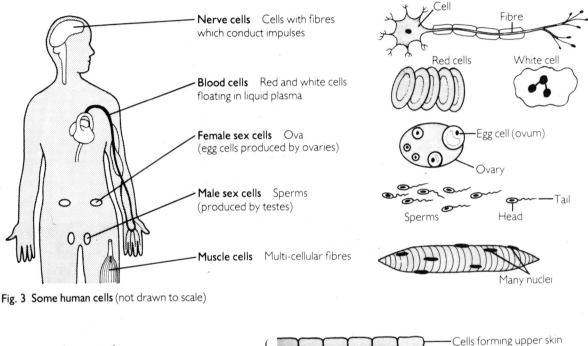

Nerve cells   Cells with fibres which conduct impulses

Blood cells   Red and white cells floating in liquid plasma

Female sex cells   Ova (egg cells produced by ovaries)

Male sex cells   Sperms (produced by testes)

Muscle cells   Multi-cellular fibres

Cell
Fibre
Red cells
White cell
Egg cell (ovum)
Ovary
Sperms
Head
Tail
Many nuclei

Fig. 3 Some human cells (not drawn to scale)

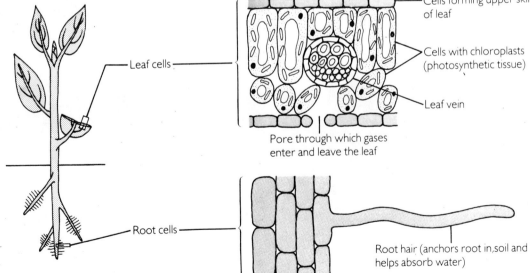

Leaf cells

Root cells

Cells forming upper skin of leaf

Cells with chloroplasts (photosynthetic tissue)

Leaf vein

Pore through which gases enter and leave the leaf

Root hair (anchors root in soil and helps absorb water)

Fig. 4 Some plant cells

1.3 The amazing body machine

The human body is a living machine, and has features which make it far superior to any machine humans have yet invented.

Growth

The human body machine makes itself. It starts as a single cell no bigger than a full stop but contains all the instructions needed to make a complete human being. This cell becomes billions of cells, each with a special job which serves the body as a whole. Unlike man-made machines the body can increase its size as it works, slowly adjusting the shape of its parts as they grow, until they are perfectly formed for the job they have to do.

Maintenance

The human body machine maintains and repairs itself. Joints are so well oiled that they work for many years without signs of wear. Parts which wear out, like red blood cells and the surface of the skin, are replaced as fast as they are lost.

Cuts which leak blood are quickly plugged and soon mend. Broken bones are mended as good as new in a few weeks.

Control

The human body machine not only controls itself but is aware of what it is doing. Apart from controlling its own movements it controls many internal processes as well such as breathing, blood flow, body temperature, and the amounts of chemicals in the blood stream. Moreover, it controls these processes without a single conscious thought.

The brain controls muscles and joints so they move in the correct order and direction, with enough speed and power. The brain also remembers what it has done, and can profit from its experience.

Fig. 1 The limits of human endurance   The longest anyone has ever lived is thought to be 116 years. A person has survived without food for 94 days, but without food *and* water for only 18 days. After breathing oxygen for 30 minutes a man survived under water for 13 minutes 42 seconds. The heaviest weight yet lifted was 2844 kg (equal to three small family cars). A technique called backlift was used. Normal weight-lifting techniques can be used to lift over 400 kg. Polynesian firewalkers can walk on glowing hot stones without damaging their feet. Nobody knows how they can do this.

Brain and nerves   The brain contains 10 000 million nerve cells and each is connected to thousands of others. The fastest nerve impulses travel at 532 k.p.h.

Ears   Human ears are sensitive to sounds from 16 cycles per second up to about 20 000 c.p.s. (the squeak of a bat). Prolonged loud noise above 140 decibels (e.g. some pop music) can permanently damage hearing.

Taste and smell   The tongue has 9000 taste buds and is sensitive to four tastes (sweet, salt, sour, and bitter) and countless flavours. The sense of smell is ten thousand times more sensitive than taste. We can smell some chemicals diluted to one part per million.

Heart   A heart beats 40 million times a year and in that time pumps 3 million litres of blood. It beats about 80 times a minute at rest rising to 200 times during strenuous exercise.

Lungs   Adult lungs have an internal area of 93 m² (40 times the external surface area of the body). You breathe about 13 500 litres of air a day.

Blood   Adults have about 5.5 litres of blood containing 30 billion red cells and 75 million white cells. Red cells live about 120 days and make 40 000 journeys around the body each month. There are 96 000 km of blood vessels in the body.

Male reproduction   Testicles produce 500 million sperms a day. After entering a woman's body a sperm must swim 146 cm to reach an egg cell (i.e. 6000 times its own length) which is equivalent to a human swimming almost 10 km.

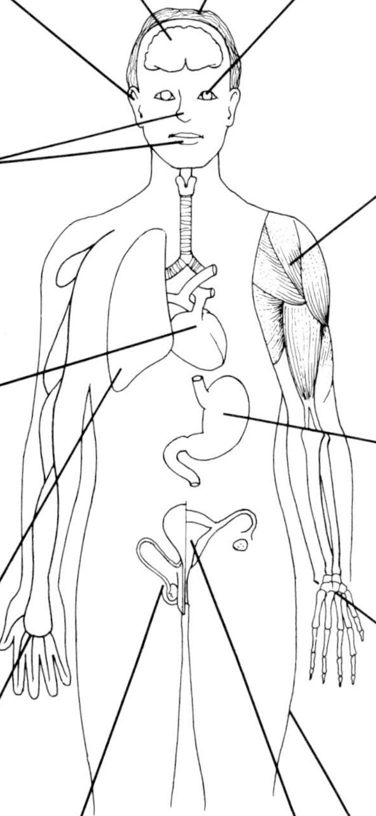

Hair   A hair grows 1 mm in three days but rarely grows more than 0.9 m long. Each hair grows for 3 to 4 years and is then pushed out by another growing from below. Humans have more body hairs than apes, but of a much shorter and softer kind.

Eyes   Each eye contains 142 million light sensitive nerve endings, 10 million of which are sensitive to colour. You can see objects as small as 0.1 mm across at a distance of 15 cm, and you can distinguish between 10 million different colour shades. You spend half an hour a day blind (the time taken up by blinking).

Muscles   You have 639 muscles which make up 40 per cent of your body weight. Well-trained athletes can produce bursts of muscular power up to 2.0 horse power lasting about ten seconds. But if you were relatively as powerful as an ant you could run at 160 k.p.h.

Food   Adults eat about 500 kg of food a year. About 11.5 litres of food and liquid (including digestive juices) pass through the gut each day.

Bones   There are 206 bones in the body. Bone is as strong as reinforced concrete and as hard as granite but far lighter than both. Bones make up only 14 per cent of body weight. An adult's thigh bone withstands 2500 kg/cm² during walking. Bone marrow produces 200 000 million red cells a day.

Skin   An adult's skin weighs about 3 kg and has an area of 1.7m². It has 8 million sweat glands and can feel vibrations with a movement as small as 0.0002 mm.

Female reproductive system   Girls have over a million egg cells in their ovaries at birth but most degenerate. The largest number of babies born at one time is ten (2 boys and 8 girls).

Fig. 2 Some amazing facts about the human body machine

1.4 Classifying and identifying organisms

More than 1 500 000 different kinds of living things have been found and more are being discovered all the time. The task of sorting all these creatures into groups is called classification.

How organisms are classified

Living things can be sorted into groups on the basis of features which are shared. In other words several different creatures can be grouped together if they have something in common.

It is very important to look not for one, but for as many shared features as possible before deciding which creatures to group together. For example, animals which fly could form a group, but this would mean including animals as different as sparrows, bats, and butterflies. As you will see in later Units, birds have more in common with fish than with butterflies.

Classification is more useful if it is based on a careful study of all the main features of organisms. Biologists have studied shapes of bodies, types of limbs, different kinds of skeletons, the arrangement of internal organs, and many other features before arriving at the modern classification system.

Biological classification

Organisms are first divided into very large groups called *Kingdoms*. Most organisms can be placed in either the *Plant Kingdom* or the *Animal Kingdom*. Unicellular organisms and fungi, which cannot be fitted into either of these groups, can be put together in the *Kingdom Protista*. Kingdoms are divided into smaller groups called *phyla*, which are divided into smaller and smaller groups. The smallest groups of living things are called *species*.

Species

A species is a group of organisms so alike that they can mate together and produce young. Humans, dogs, dandelions, and cowslips are examples of four different species. Usually members of one species cannot breed with members of another. Different types of dog can mate and produce puppies, but dogs and cats cannot breed together.

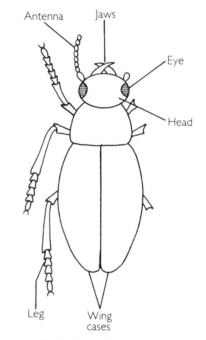

Fig. 1 Diagram of a beetle (right-hand legs omitted). Study this drawing before using the key illustrated in Figure 2.

A
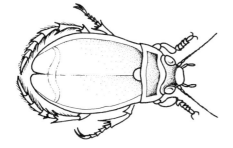

B
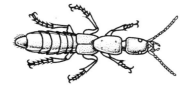

C
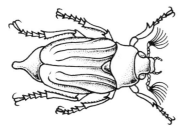

D
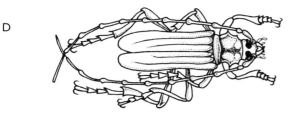

E
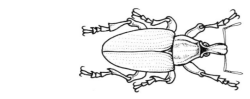

F
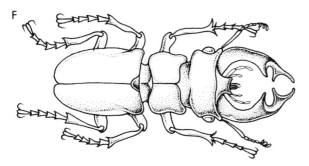

Fig. 2 A simple key

Identifying organisms with a key

If you wish to find out which species an organism belongs to, or its common name, a device called a *key* can be helpful. One of the commonest and simplest types of key consists of brief descriptions arranged in numbered pairs. Figure 2 is an example of this type of key. Use the key in the following way.

First study Figure 1 and learn the technical names for the parts of a beetle. Then look at beetle **A** in Figure 2 and read the first pair of descriptions. Decide which description fits the beetle. Opposite the description you choose you will find either the name of the beetle or a number. If you find a number, it tells you which pair of descriptions to read next. Read them and decide which description fits beetle **A**. Opposite the description you choose you will again find either the name of the beetle or the number of the next pair of descriptions to read. Repeat this process until you find the name of the beetle. Then start all over again with beetle **B**. After you have repeated the process for each beetle in turn, you will have discovered the name of each one. Do *not* write in the text book.

Key

1	Hind legs pointed and fringed with hairs	Diving beetle
	All legs in two claws	2
2	Wing cases very small	Rove beetle
	Wing cases large	3
3	Large jaws	Stag beetle
	Jaws small, or jaws not shown in drawing	4
4	Antennae have brush-like tufts of hairs at tip	Cockchafer
	Antennae not like above	5
5	Antennae longer than body	Long-horned beetle
	Antennae shorter than body	Weevil

Exercises

1   What does 'classification' mean? What do scientists look for when deciding how to classify organisms?

2   Name the three major groups in a system of biological classification. Name the smallest group in such a system.

3   What determines whether two organisms are classified as members of the same species?

9

1.5 Protists: protozoa, fungi, and algae

Protists are organisms which are hardly ever noticed and yet they live all around us in countless millions. Compared with animals and plants protists have a very simple structure. Many consist of one cell and are too small to see without a microscope. Others are multicellular and have groups of cells specialized into tissues.
(Exercises for Units 1.5–1.11 are on page 26.)

Protozoa

Protozoa are microscopic, unicellular protists (Fig. 3). Most feed by catching other protists, but some live as parasites in the bodies of larger organisms, causing diseases such as malaria and dysentery. Others, like *Euglena*, have chlorophyll and feed by photosynthesis.

Rhizopods   These are protozoa which move and usually feed by pushing out parts of their bodies to form temporary arms called **pseudopodia** (Figs. 1 and 3A).

Ciliates   These are protozoa with microscopic hairs called **cilia**, which they use for feeding and moving (Figs. 2 and 3B).

Flagellates   These are protozoa which move about by lashing a whip-like hair called a **flagellum** (Fig. 3D). Many have chlorophyll.

Fungi

Moulds, mushrooms, and yeasts are examples of fungi. Most fungi are multicellular, although yeasts consist of one cell (Fig. 4B). Multicellular fungi are usually made up of fine threads which either spread out in a thin network like bread mould (Fig. 4A) or are bundled together and form more solid structures, like mushrooms (Fig. 4C).

Yeasts feed by changing sugar into alcohol and carbon dioxide gas. This is called **fermentation** and is used in brewing, wine making, and baking.

Most moulds, mushrooms, and toadstools are **saprophytes**. This means they feed by decomposing dead organisms, and substances such as stored food and wood which were once part of living organisms.

Some fungi live as parasites, causing diseases like ringworm in humans and mildew in plants.

Algae

Algae are simple plant-like organisms (Fig. 5). They live by photosynthesis. Green algae live in the sea, in fresh water, or in damp places on land. They occur in several forms: as single cells, hollow balls of cells, fine threads, or hollow tubes. Brown algae live in the sea where they often reach several metres in length.

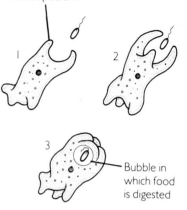

Pseudopodium

Bubble in which food is digested

Fig. 1 Amoeba feeding *Amoeba* (and other Rhizopods) move and feed by forming temporary arms (pseudopodia). These surround food, enclosing it in the amoeba's body where it is digested.

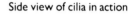

Side view of cilia in action

Power stroke   Recovery stroke

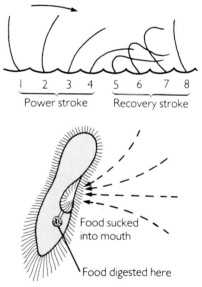

Food sucked into mouth

Food digested here

Fig. 2 Paramecium feeding *Paramecium* (and other ciliates) have tiny hairs called cilia. These flick back and forth like oars, rowing *Paramecium* through the water, and causing currents of water which suck food into its mouth.

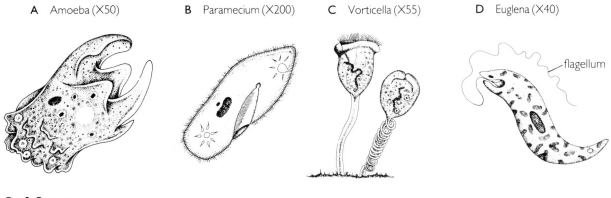

A Amoeba (×50)   **B** Paramecium (×200)   **C** Vorticella (×55)   **D** Euglena (×40)

flagellum

Fig. 3 Protozoa

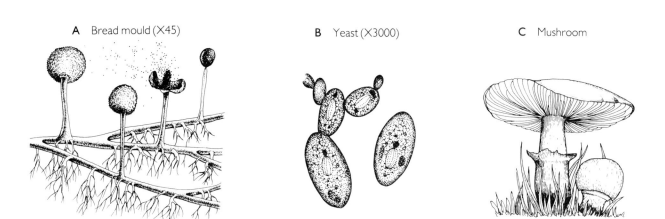

A Bread mould (×45)   **B** Yeast (×3000)   **C** Mushroom

Fig. 4 Fungi

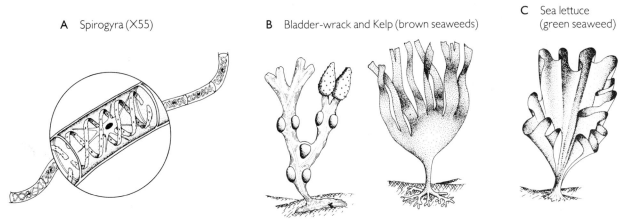

A Spirogyra (×55)   **B** Bladder-wrack and Kelp (brown seaweeds)   **C** Sea lettuce (green seaweed)

Fig. 5 Algae

1.6 Coelenterates, flatworms, and true worms

This unit and the next five describe the Animal Kingdom. Animals are killers and thieves: to get food they must either steal it (e.g. fruits, seeds) or kill it. Animals which have a backbone (vertebral column) are called vertebrates, and those without a backbone are invertebrates. This Unit and the next two describe invertebrate animals.

Coelenterates

Sea anemones and jellyfish are coelenterates found in the sea, and *Hydra* is an example found in ponds and streams (Fig. 3). Coelenterates have a body like a bag: the opening at one end is the mouth and it is usually surrounded by tentacles. The walls of the bag consist of two layers of cells (Fig. 1). The tentacles are armed with weapons called sting cells. When its trigger is touched a sting cell shoots out a long thin thread. Some threads inject paralysing poison into the prey.

Flatworms (Platyhelminths)

Flatworms have flat bodies with a mouth at one end. *Planarians* are flatworms which live in fresh water and in damp soil (Fig. 4). They swim or crawl by using the cilia covering their bodies. Tapeworms and liver flukes live as parasites inside the bodies of animals including humans (Fig. 4).

True worms (segmented, or annelid worms)

The body of a true worm consists of similar compartments called **segments**. The segments are marked off by rings around the body (Figs. 2 and 5). Earthworms live in the soil feeding on the remains of dead plants and animals. Peacock worms live in tubes in sand and use a fan of long bristles around the mouth to filter tiny organisms from sea water. Leeches live mostly as parasites, sucking blood from their victims.

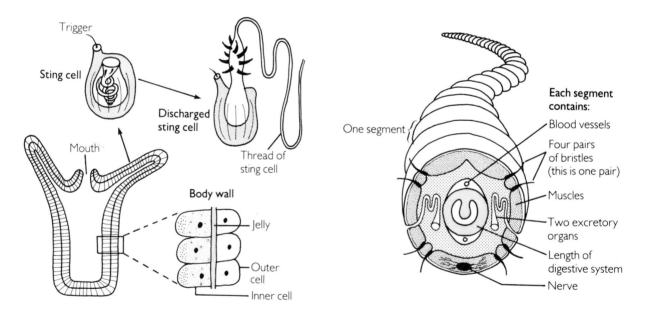

Fig. 1 Hydra is a typical coelenterate It consists of two layers of cells with jelly inbetween. Tentacles have sting cells used to catch prey and for defence.

Fig. 2 The earthworm, showing how the body consists of compartments called segments.

Hydra

Sea anemone

Jellyfish

Fig. 3 Coelenterates

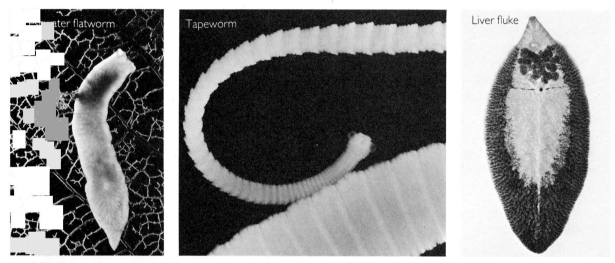

...ter flatworm

Tapeworm

Liver fluke

Fig. 4 Flatworms (platyhelminths)

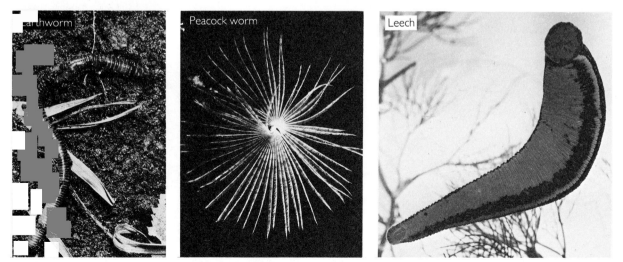

...rthworm

Peacock worm

Leech

Fig. 5 True worms (segmented or annelid)

1.7 Arthropods

Imagine creatures with skin as tough as leather, with up to 30 000 lenses in each eye, which taste with the soles of their feet, and which have feelers (antennae) that can smell something 10 kilometres away. These are just a few features of arthropods.

What is an arthropod?

There are more species of arthropod than of any other type of animal (Fig. 1). Arthropods have segmented bodies and are encased in a tough outer skin, the *cuticle*, which protects and supports them. The cuticle can be rigid like the 'shell' of a crab, or leathery and flexible as in insects. The cuticle is always thin and flexible where the body and limbs bend (Fig. 2). The head of an arthropod carries sense organs. Most have antennae which are sensitive to touch, temperature, sound, taste, and smell. Arthropod eyes are sometimes made up of thousands of tiny visual units. Such eyes are called *compound eyes* (Fig. 3).

Crustaceans (Fig. 4)   Crabs and lobsters are crustaceans with a hard chalky cuticle. Other types, like water fleas, have a thin cuticle. Crustaceans have two pairs of antennae, and between five and nineteen pairs of legs.

Arachnids (Fig. 5)   This group includes spiders, harvestmen, scorpions, mites, and ticks. Arachnids have four pairs of legs and no antennae. Their bodies are divided into two parts and the head has pincer-like jaws.

Insects (Fig. 6)   Several million types of insect are known. They have three pairs of legs and many have wings. Their bodies consist of three parts: a head with eyes and antennae, a thorax with wings and legs, and an abdomen (Fig. 3).

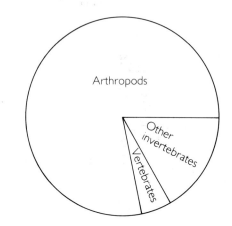

Fig. 1 A pie-chart showing that there are more species of arthropod than any other type of animal. Millions of species are already known, and it is certain there are millions more yet to be discovered.

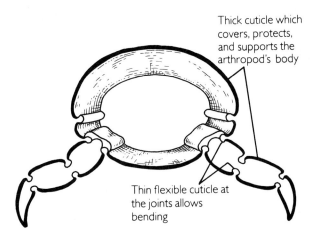

Fig. 2 Arthropods are supported by a tough outer skin called the cuticle.

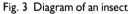

Fig. 3 Diagram of an insect

Fig. 4 Crustaceans

Fig. 5 Arachnids

Fig. 6 Insects

1.8 Myriapods, molluscs, and echinoderms

Do centipedes have a hundred feet, and how many do millipedes have?
Molluscs have only one foot but echinoderms have hundreds, each ending in a
tiny sucker.

Myriapods

This group (Fig. 3) includes centipedes and millipedes. Myriapods are arthropods with long bodies consisting of many similar segments.

Centipedes   These have one pair of legs on each segment. The shortest centipede is only 5 mm long, and has sixteen legs. The longest is up to 330 mm long and has 177 pairs of legs. Centipedes are carnivorous and paralyse their prey with poison fangs.

Millipedes   Unlike centipedes, millipedes have two pairs of legs per segment and most eat plants. One species has 355 pairs of legs.

Molluscs

These are not segmented. They have soft bodies, usually protected by one or two shells (Fig. 4). Limpets and snails both have one shell and move about on a large slimy foot. They feed with a long tongue called a *radula*. This tongue has a surface as rough as a file and is used to scrape pieces off plants, or microscopic creatures off rocks (Fig. 1). Mussels and oysters have two shells hinged together. In slugs and cuttlefish the shell is inside the body, while the

octopus has no shell. Cuttlefish, squid, and octopus use arms with suckers to catch their prey. Then they tear it to pieces with a radula and a beak curved like a parrot's (Fig. 2). Cuttlefish, squid, and octopus are jet propelled, and discourage attackers by squirting ink at them.

Echinoderms

This group (Fig. 5) includes starfish, sea urchins, and sea cucumbers. They have armour-plated skin covered with spines. There is no head region and the body is often formed of parts which radiate outwards, like the spokes of a wheel (e.g. starfish). Echinoderms move slowly on tubular feet with suckers at the ends.

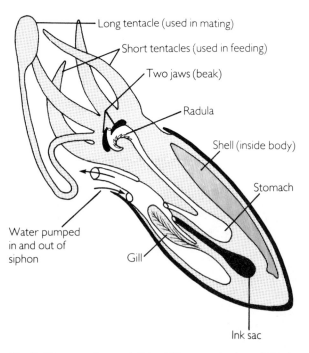

Long tentacle (used in mating)
Short tentacles (used in feeding)
Two jaws (beak)
Radula
Shell (inside body)
Stomach
Water pumped in and out of siphon
Gill
Ink sac

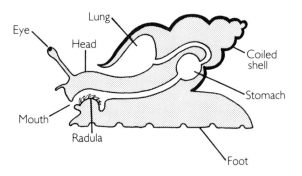

Eye
Lung
Head
Coiled shell
Stomach
Mouth
Radula
Foot

Fig. 1 Diagram of a snail   The radula is used like a file to scrape up food. A snail moves by making ripples pass along its foot from the hind end to the front.

Fig. 2 Diagram of a cuttlefish   When attacked, cuttlefish become jet propelled by squirting water out of their mantle cavity and through the siphon. They also release a cloud of ink to confuse the attacker.

Centipede

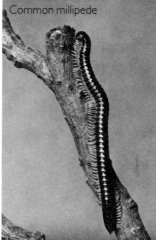
Common millipede

Flat-backed millipede

Fig. 3 Myriapods

Snail

Mussels

Cuttlefish

Fig. 4 Molluscs

Starfish

Sea cucumber

Sea urchin

Fig. 5 Echinoderms

1.9   Fish, amphibia, and reptiles

Did you know that before you were born (when you were only 2 mm long) you had gill slits like a fish, and a tail? These features soon disappeared but the fact that they existed at all, plus the fact that you have a backbone (vertebral column), show that you are a vertebrate. This Unit and the next describe vertebrate animals (Fig. 2).

Fish

Fish (Fig. 3) have features which make them perfectly adapted for life in water. They have a streamlined shape, are covered with overlapping scales, breathe through gills, and move using fins. Fish such as sharks have a skeleton of gristle (cartilage) but the majority of fish have a skeleton of bone. Bony fish, like cod and haddock, are weightless in water because they are buoyed up by an air-filled space in their bodies called a **swim bladder** (Fig. 1).

Amphibia

These animals include frogs, toads, newts, and salamanders (Fig. 4). They have moist skin without scales. They can live under water breathing through their skin, but when on land they breathe with lungs. They lay their eggs in water, and these hatch into larvae called tadpoles which breathe with gills and swim with a tail.

Reptiles

Reptiles (Fig. 5) have a dry scaly skin. Even the types which live in water breathe with lungs. Reptile eggs do not have to be laid in water because they have tough leathery coverings which stop them drying up. The eggs do not hatch into larvae, but into small versions of their parents.

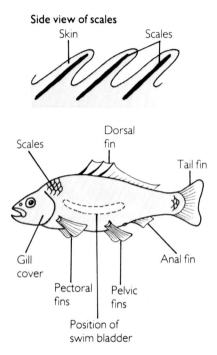

Side view of scales

Fig. 1 Bony fish have scales which overlap like roof tiles, so water flows smoothly over them. The swim bladder is filled with air until the fish is weightless, so it does not have to waste energy stopping itself from sinking.

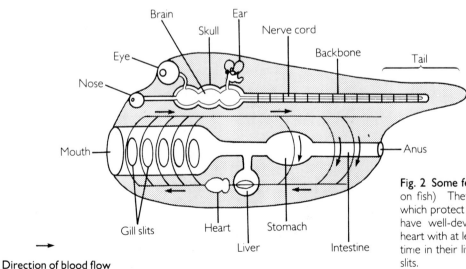

Direction of blood flow

Fig. 2 Some features of vertebrates (based on fish) They have a backbone and skull which protect a brain and nerve cord. They have well-developed sense organs and a heart with at least three chambers. At some time in their lives they all have a tail and gill slits.

Shark Fish Sea horse

Fig. 3 Fish

Frog Toad Newt

Fig. 4 Amphibia

Tortoise Crocodile Snake

Fig. 5 Reptiles

1.10 Birds and mammals

Which birds can't fly, and which mammals lay eggs?

Birds

In birds which can fly, almost every part of the body is adapted to help them fly efficiently. They have a streamlined shape (even the ears are covered with feathers) and their bones are either hollow, honeycombed, or moulded into thin sheets for lightness.

Birds feed using a light beak with no teeth. Their lungs extend into large air sacs which lighten the body and assist in breathing. The heart is very large—up to one-tenth of body weight. It beats fast during flight, circulating blood quickly to the flight muscles. The flight muscles are also large and powerful—up to one-fifth of body weight. They are attached to a huge breast bone which sticks out from the chest like the keel of a boat (Fig. 1).

Flight feathers on the wings and tail are responsible for flight; *contour feathers* cover the body giving it a streamlined shape; and fluffy *down feathers* keep the body warm (Fig. 2).

The kiwi, ostrich, and penguin are examples of flightless birds. All birds are warm-blooded, which means they have a constant body temperature (Fig. 3).

Mammals

Mammals, like birds, have a constant body temperature, and some of the warmth is kept in by their hairy skin (Fig. 4). Female mammals suckle their young on milk from mammary glands (breasts).

Monotremes   These are egg-laying mammals, and include the spiny anteater and duck-billed platypus. After hatching, the young are fed on milk from teatless breasts.

Marsupials   These are the pouched mammals and include opossums, koalas, bandicoots, and kangaroos. Young marsupials are born before they are completely developed. After birth they crawl up to their mother's pouch where they obtain milk from a teat and complete their development.

Placental mammals   This is the largest group of mammals and includes humans. Their young develop to an advanced stage before they are born. Development takes place inside the mother's body where they are attached to the wall of the womb by an organ called the *placenta*. The placenta supplies a developing baby with food and oxygen.

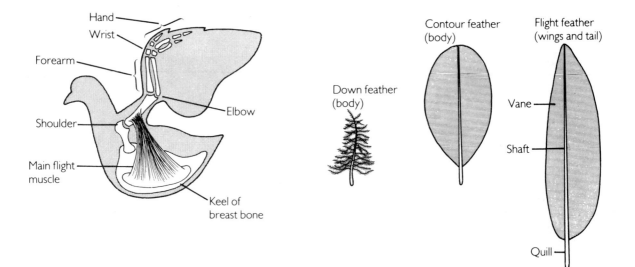

Fig. 1 **The largest flight muscles of a bird** are those which pull the wings downwards. These muscles pull against a keel-shaped extension of the breast bone.

Fig. 2 **Feathers** are light but strong. A bird preens (cleans) its feathers with its beak, and some birds waterproof their feathers with oil from a preen gland at the base of the tail.

Fig. 3 Birds

Fig. 4 Mammals

1.11 Plants

Would you like to sit in the sun all day using its light energy to make your food from water and air? Plants do this all their lives. A few get extra food by catching insects and some live as parasites sucking juices from other plants. The smallest plants are mosses and duckweeds less than a millimetre across, and the largest, the General Sherman tree, is a giant redwood in the Sequoia National Park in California. It is 83 metres high and the largest living thing on Earth.

Mosses and liverworts (Bryophytes)

These plants do not have proper roots, stems, or leaves but often have structures which look like them (Fig. 1).

Mosses and liverworts require water for reproduction because male plants produce sperms which must swim to a female plant to fertilize it. After fertilization a female plant grows a capsule which releases microscopic spores. The spores grow into new plants.

Ferns, horsetails, and club mosses (Pteridophytes)

These plants have true roots, stems, and leaves, but no flowers (Fig. 2).

They reproduce by means of spores from tiny capsules clustered on the under-side of their leaves. Spores grow into a plant more like a liverwort than a fern. These tiny plants produce sperms and eggs and after fertilization an egg grows into a new fern plant.

Seed plants

These have reproductive organs which produce seeds.

Cone-bearing plants (Gymnosperms) This group of plants includes firs, pines, and spruces. Their seeds are produced inside cones.

Flowering plants (Angiosperms) These plants have flowers containing reproductive organs which produce the seeds. Male organs called **stamens** produce pollen grains. These fertilize ovules (eggs) in female organs called **carpels**. A fertilized ovule develops into a seed contained in a fruit. **Monocotyledons** are flowering plants whose seeds germinate into a seedling with only one seed leaf (cotyledon). Their leaves have parallel veins. This group includes grass, tulip, daffodil, and lily (Fig. 3). **Dicotyledons** are the largest group of flowering plants. Their seedlings have two seed leaves and their leaves have branching veins (Fig. 4).

Marchantia, a common liverwort

White fork moss

Fig. 1 Mosses and liverworts

Fig. 2 Ferns

Fig. 3 Monocotyledons

Fig. 4 Dicotyledons

1.12 Evolution

Some people believe the human race was specially created a few thousand years ago, and that other living things were separately created. There is no scientific evidence to support or disprove this idea. The theory of evolution on the other hand, says that all living things were produced over billions of years by gradual random changes from creatures that lived in the past. There is an enormous amount of scientific evidence which supports this theory, but the most convincing evidence comes from a study of fossils.

Evolutionary change

Fossil collections show that in the 2 or 3 billion years since life first appeared on earth, there has been an increase in the number of different organisms, and a development from simple to complex organisms (Fig. 1). How could this have happened?

Survival of the fittest

No two organisms are exactly alike—even 'identical' twins differ slightly. If an organism is born with differences which make it better adapted than its fellows to its particular way of life, it will have a better chance than its fellows of survival, and of producing young which will survive.

An organism's chances of survival could be increased because it is better equipped to find food, water, a mate, or shelter; or to fight off predators; or to withstand heat, cold, drought, diseases, etc.

The fact that better-adapted organisms are more likely to survive than their less well-adapted fellows is called the *survival of the fittest*.

Descendants of well-adapted organisms may produce young with different adaptations which help them to survive. Over billions of years this process could give rise to an entirely new type of organism (species). This is how life could have evolved from simple creatures found in the oldest rocks to the millions of different creatures alive today.

Did humans evolve from apes?

Humans definitely did *not* evolve from any of the apes alive today. But humans and modern apes almost certain did evolve from a shared ape-like ancestor.

Fossils show that this ancestor—neither ape nor human—lived about 50 million years ago in East Africa. Fossils from younger rocks seem to show that descendants of this animal split into two separate groups: one of which evolved into apes, and the other into humans.

It is possible to guess what a creature looked like from only a few bits of its fossilized skeleton. This has enabled scientists and artists to produce the drawings of our ancestors included in Figure 2.

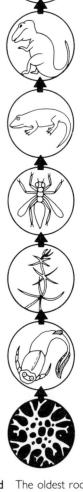

First flowering plants
(75 million years ago)

First mammals
(135 million years ago)

First seed plants
(230 million years ago)

First reptiles
(250 million years ago)

First amphibia
(325 million years ago)

First insects
(350 million years ago)

First land plants
(360 million years ago)

First fish
(400 million years ago)

First life
(3 billion years ago)

Fig. 1 The fossil record The oldest rocks contain no trace of life. Younger rocks above them contain fossils of increasing numbers of different organisms, and a development from simple to complex forms.

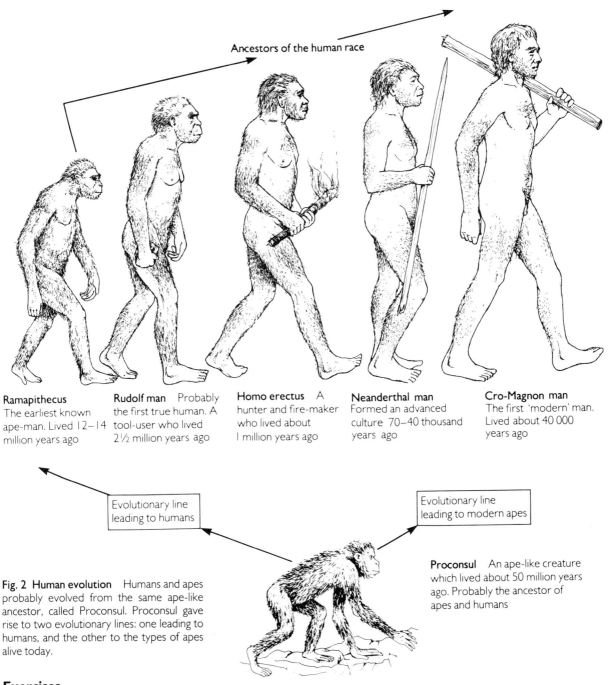

Ramapithecus
The earliest known ape-man. Lived 12–14 million years ago

Rudolf man Probably the first true human. A tool-user who lived 2½ million years ago

Homo erectus A hunter and fire-maker who lived about 1 million years ago

Neanderthal man Formed an advanced culture 70–40 thousand years ago

Cro-Magnon man The first 'modern' man. Lived about 40 000 years ago

Ancestors of the human race

Evolutionary line leading to humans

Evolutionary line leading to modern apes

Fig. 2 Human evolution Humans and apes probably evolved from the same ape-like ancestor, called Proconsul. Proconsul gave rise to two evolutionary lines: one leading to humans, and the other to the types of apes alive today.

Proconsul An ape-like creature which lived about 50 million years ago. Probably the ancestor of apes and humans

Exercises

1 What provides the best scientific evidence to support the theory of evolution?

2 Write a paragraph to explain what is meant by 'survival of the fittest'.

3 How long did it take for the first 'modern' man to evolve from the first ape-man?

4 What is the earliest known ape-man called?

5 What is the name of the ancestor shared by humans and apes? When did this ancestor live?

6 When did the first fish evolve? How much longer did it take for the first mammals to evolve?

Topic 1 exercises

1   What is a cell, and what are the main differences between animal and plant cells?

2   What are the functions of the cell membrane and the nucleus of a cell?

3   Match the following parts of an organism with the organisms which possess them:

Parts: pseudopodia, cilia, flagella, antennae, radula, sting cells, tube feet, poison fangs, swim bladder, mammary glands

Organisms: snail, *Amoeba*, starfish, women, *Paramecium*, mackerel, *Euglena*, beetle, sea anemone, centipede.

4   What are tissues and what are organs? Name one example of each.

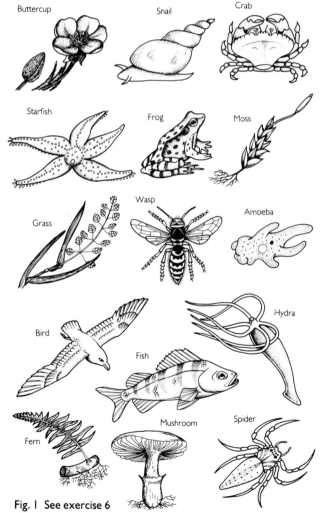

Fig. 1 See exercise 6

5   What is meant by the word species? Which of the following are species: dandelions, trees, humans, amoebas, birds, dogs, mosses?

6   a) Arrange the organisms in Figure 1 into plants and animals.
b) Arrange the plants into those which produce seeds and those which don't. Arrange the seed plants into dicotyledons and monocotyledons.
c) Arrange the animals into invertebrates and vertebrates, and list the features of vertebrates. Which of the vertebrates is warm-blooded?
d) From among the invertebrates find a coelenterate, arachnid, mollusc, insect, crustacean, and echinoderm.
e) Which of the organisms in Figure 1 are segmented; which produce spores; and which have chlorophyll?

7   What determines whether an animal is called a vertebrate or an invertebrate?

8   Arthropods and true worms have segmented bodies, but are not classified together. Name two reasons why.

9   How many cells do protozoa have? Name three different types of protozoa.

10   Which animals can breathe both on land and in water? Explain how they are able to do this.

11   Name one mollusc whose shell is inside its body.

12   What are pseudopodia, flagella, and cilia used for?

13   What is a larva?

14   A reptile and an amphibian both lay eggs. Describe how the eggs are different, and how the animals which hatch from them do or do not resemble their parents.

15   Name four major groups of arthropods.

16   List all the features of birds which are concerned with flight.

17   Plants do not have to find food; they make it. What three things do they need to make food?

18   What are the segments of an insect's body called?

19   Name one arthropod with 8 legs; one with 6 legs; and one with a hard, chalky cuticle.

20   Of what use to a fish is its swim bladder?

21   a) Describe the different ways used by each group of mammals to produce their young.
b) All mammals suckle their young. What does this mean?

22   What is the largest group of flowering plants called?

26

Living things need each other

The sea anemone and hermit crab help each other by living together

2.1 How plants make food

Plants use carbon dioxide, water, and sunlight to make food which is used not only by themselves, but by all other life on Earth. In addition, this food-making process creates the oxygen which makes air breathable.

What is photosynthesis?

Unlike animals, plants do not eat other living things. They make their food by a process called **photosynthesis**. This takes place in the leaves and other green parts of a plant.

Photosynthesis produces glucose sugar. This sugar is transported to all parts of a plant where some is respired to produce energy, some is changed into starch and stored for future use, and some is used in combination with chemicals absorbed from the soil to manufacture all the substances in the body of a plant.

Leaves are food factories

To make food, plants require energy, chlorophyll, carbon dioxide gas, and water (Fig. 1). The energy comes from sunlight and is absorbed by chlorophyll—the green substance found in leaves (Fig. 2B). Carbon dioxide comes from the air and enters a plant through tiny pores called **stomata** (singular: stoma) (Fig. 2C). Water is absorbed from the soil by roots. It travels through tiny tubes upwards through the plant to the leaves, where the tubes form a branching network of veins (Fig. 2A).

Chlorophyll absorbs light energy and uses it to combine carbon dioxide with water. This produces glucose sugar, and releases oxygen gas as a by-product. The oxygen passes out through stomata into the air where it is used in plant and animal respiration.

Underside of a leaf showing veins

Exercises

1 What substances does a plant use to make food by photosynthesis, and what food does this process make?

2 What parts do light and chlorophyll play in photosynthesis?

3 What gas enters stomata during photosynthesis and what gas passes out of the stomata during photosynthesis?

4 If there were no photosynthesis, the air would quickly become unbreathable. Why is this so?

5 What is a stoma?

6 Where is chlorophyll found in a plant?

7 How is a leaf like a sandwich (Fig. 2)?

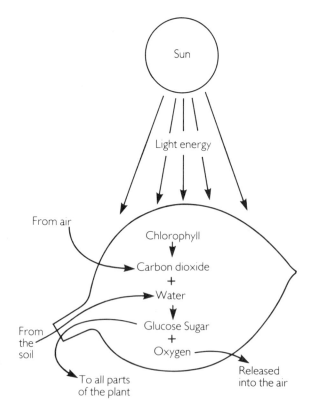

Fig. 1 Summary of photosynthesis

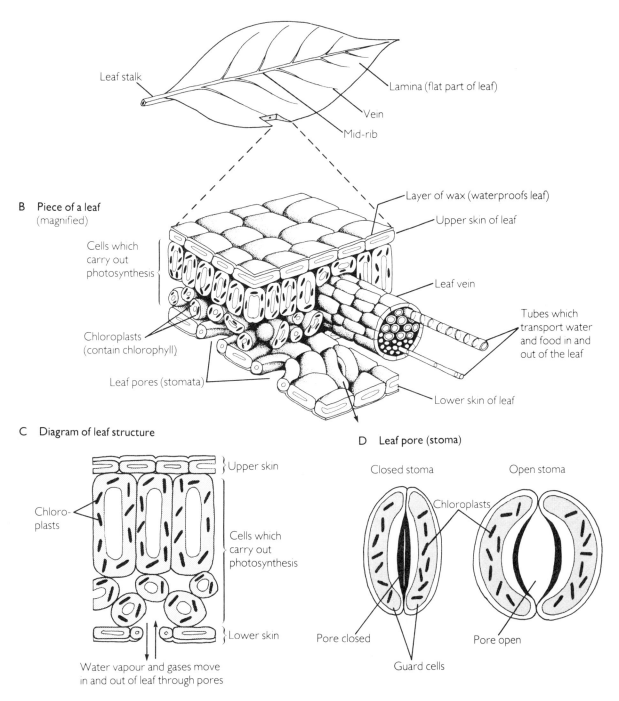

A   Parts of a leaf

Leaf stalk

Lamina (flat part of leaf)

Vein

Mid-rib

B   Piece of a leaf (magnified)

Layer of wax (waterproofs leaf)

Upper skin of leaf

Cells which carry out photosynthesis

Leaf vein

Tubes which transport water and food in and out of the leaf

Chloroplasts (contain chlorophyll)

Leaf pores (stomata)

Lower skin of leaf

C   Diagram of leaf structure

Chloro-plasts

Upper skin

Cells which carry out photosynthesis

Lower skin

Water vapour and gases move in and out of leaf through pores

D   Leaf pore (stoma)

Closed stoma

Open stoma

Chloroplasts

Pore closed

Pore open

Guard cells

Fig. 2 **Structure of a leaf**   A leaf is like a sandwich: it consists of an upper and lower skin with photosynthetic tissue inbetween. In most plants the lower skin of a leaf has pores called stomata.

2.2 Food chains and food webs

All life on Earth depends upon sunlight.

Sunlight—the energy of life

If green plants could not use sunlight energy to make food by photosynthesis, practically every living thing on Earth would die. This would happen for two reasons: first, photosynthesis releases the oxygen which keeps air breathable; and second, photosynthesis produces food.

Plants are almost the only living things which can make their own food. They are therefore called the **producers** of the world.

Nearly all other creatures depend directly or indirectly upon plant food and so they are called **consumers**. Herbivores, such as cattle and rabbits, are called **primary consumers** because they depend entirely upon plant food. Carnivores, like foxes and hawks, are called **secondary consumers** because they eat herbivores and each other, and so depend *indirectly* on plants for food.

Put simply, this means that sunlight provides energy which keeps all Earth's creatures alive. Producers (green plants) convert sunlight energy into food, and consumers share this energy among themselves when they eat plants and each other. This flow of energy from producers to consumers leads to the formation of food chains and food webs.

Food chains and webs

Plants → rabbits → humans is a **food chain** which occurs when rabbits eat grass and humans eat the rabbits. All food chains begin with green plants because these are producers, and the following 'links' in the chain are consumers. A simple food chain is shown in Figure 1.

Food chains are rarely as simple as this, because consumers usually eat more than one food, and may themselves be eaten by several consumers. Humans, for instance, eat foods other than rabbit, and rabbits are also eaten by foxes and hawks. In this way, several food chains become connected together making what is called a **food web**. Figure 2 illustrates part of a food web which exists in a pond.

Parasites, scavengers, and decomposers

Carnivores are not the only consumers:

Parasites consume food from all parts of a food web (e.g. mildew fungi feed off plants, and fleas feed off animals).

Scavengers, like carrion crows, eat dead organisms.

Decomposers, such as bacteria and fungi, play a vital part in food webs by decomposing dead things into a liquid which keeps soil fertile and ensures healthy plants.

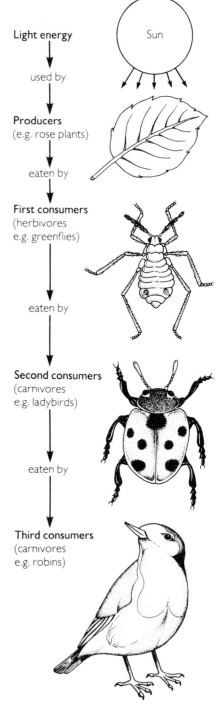

Light energy
↓ used by

Producers
(e.g. rose plants)
↓ eaten by

First consumers
(herbivores
e.g. greenflies)
↓ eaten by

Second consumers
(carnivores
e.g. ladybirds)
↓ eaten by

Third consumers
(carnivores
e.g. robins)

Sun

Fig. 1 A simple food chain

Exercises

1 Why are green plants known as producers, and why are most other creatures known as consumers?

2 Sort the following into producers, primary and secondary consumers, parasites, scavengers, and decomposers: cow, buttercup, lion, bread mould, lettuce, earthworm, bed bug, caterpillar, tapeworm, carrion crow, mushroom, fox, leech.

3 Write out each list to form a food chain:
   a) caterpillar, cabbage, hawk, blue-tit
   b) duck, pondweed, water snail, large water beetle
   c) eagle, grass, stoat, rabbit.

4 Study Figure 2 and then:
   a) write out two food chains ending in a fish louse
   b) write out two food chains ending in a leech
   c) write out one food chain ending in a water beetle.

5 a) List the herbivores, carnivores, and parasites in Figure 2.
   b) List the first and second consumers in Figure 2.

6 What would happen to all the other organisms in the pond if water beetles were removed?

7 Name the decomposers in Figure 2. How do they help maintain the pond community?

8 Name the scavengers in Figure 2. What part do they play in maintaining the pond community?

9 Which organisms, other than frogs, would suffer if children removed all the tadpoles from a pond?

Food web in a pond

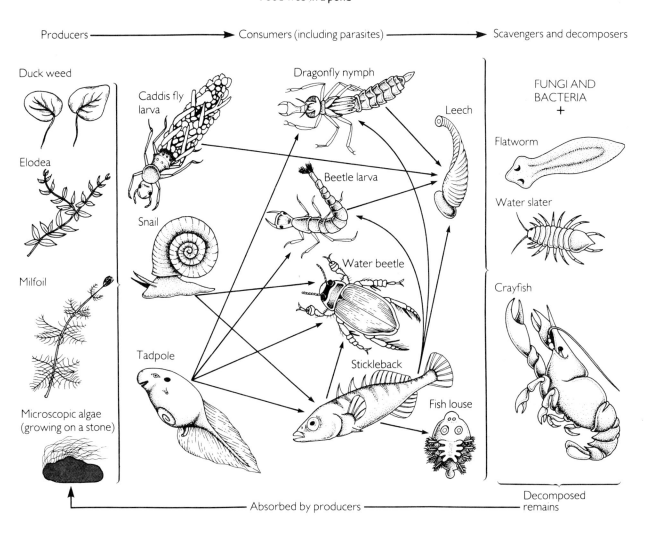

Producers ──────────────→ Consumers (including parasites) ──────────────→ Scavengers and decomposers

Duck weed

Elodea

Milfoil

Microscopic algae (growing on a stone)

Caddis fly larva

Snail

Tadpole

Dragonfly nymph

Beetle larva

Water beetle

Stickleback

Leech

Fish louse

FUNGI AND BACTERIA
+
Flatworm

Water slater

Crayfish

── Absorbed by producers ──

Decomposed remains

Fig. 2 A food web consists of several food chains.

2.3 Studying the environment

The world is a ball of rock covered in places with water or soil, and enclosed in an envelope of air. This is the environment we share with countless millions of other creatures. This worldwide environment is divided into small areas called habitats, each with its own community of living things.

Habitats

A *habitat* is any place where a group of organisms can live. Ponds, streams, rock pools, moorlands are all habitats. Large habitats, like forests, are made up of many smaller habitats. Each type of tree in a forest is a separate habitat (Fig. 2), and the carpet of dead leaves on a forest floor is another.

Communities

A *community* is the group of organisms which live in a habitat (Fig. 1). A rock pool, for instance, has a community which includes sea weeds, starfish, crabs, and sea anemones.

All the organisms in a community depend upon their habitat and upon each other for a living, and they are linked by food chains which, together, form a complicated food web.

Ecosystems

An *ecosystem* is the name for a habitat *and* the community of organisms which live in it. For example, the water, mud, and rocks of a pond together with all the organisms which live in it make up a pond ecosystem. *Ecology* is the scientific study of ecosystems.

Exercises

1   What is the difference between a community and a habitat?

2   What is an ecosystem? Describe an example. What does an ecologist do?

3   Study a community, using Figures 1 and 2 as a guide. List some food chains in the community.

4   Study the organisms illustrated in Figure 2. Which of them are primary consumers, and which could be secondary consumers? Which uses the tree only as a place to grow and shelter?

A rotting log community   Ants, spiders, beetles, moss, fungi, etc., live in rotting wood. Keep the log moist. A ring of vaseline around the upper edge of the glass will stop insects escaping. The case can be made from glass or perspex taped together and placed in a tray. Feed ants with crumbs placed on a sponge moistened with sweetened water.

A pond community   A washing-up bowl will do but an aquarium is best. Spread washed gravel on the bottom sloping it up to the back. Plant different types of pond weed and fill with clean pond water. Wait a week before introducing animals. Do not introduce dragonfly larvae or beetle nymphs as these will eat everything else. A freshwater mussel will filter the water. Avoid full sunlight as this will over-heat the water and turn it green.

Fig. 1 Communities you can study at home or in school.

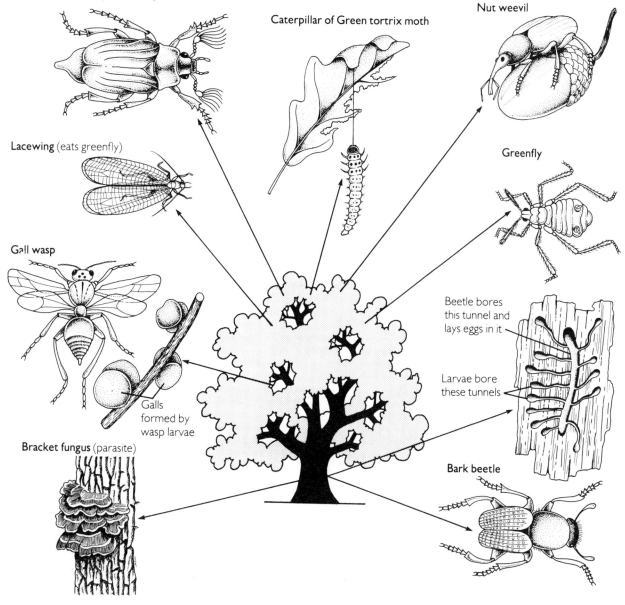

Cockchafer beetle (eats leaves)

Caterpillar of Green tortrix moth

Nut weevil

Lacewing (eats greenfly)

Greenfly

Gall wasp

Beetle bores this tunnel and lays eggs in it

Larvae bore these tunnels

Galls formed by wasp larvae

Bracket fungus (parasite)

Bark beetle

Fig. 2 An oak tree community This is a community which obtains food or shelter (or both) from the tree. Each type of tree has a different community, and a different food web. You can discover much for yourself about such a community by visiting one particular tree regularly throughout the year. Examine its bark (without damaging it) for fungi, moss, lichens, insects, centipedes, spiders, etc. Search the leaves for galls, caterpillars, aphids, leaf miners, and other creatures which eat foliage. Make notes, and distinguish between temporary visitors to the tree and those which are part of its community.

2.4 Soil

Soil is not as dead as it looks. A handful of garden soil contains millions of bacteria, protozoa, and microscopic fungi, while a cubic metre of woodland soil and leaf litter can contain as many as 250 000 creatures.

What is soil made of?

Soil consists of tiny fragments of rock, with humus, water, air, and living organisms (Figs. 1 and 2). *Humus* is the decaying remains of dead plants and animals.

Rock particles

These come from rocks which have been broken down over thousands of years by rain water, frost, and pressure from growing plant roots. Rain contains chemicals which dissolve some rocks and leave others behind in tiny pieces. Frost breaks up rock by causing water inside it to expand. When rocks are washed into rivers they are rolled over and over and pounded together. This breaks them into powder called *alluvium* which is spread over flat areas of land where rivers reach the sea.

The smallest rock particles in soil are called clay, the next largest are silt, then comes sand, and the largest particles are called gravel (Fig. 1).

Humus

When animals and plants die their remains are decomposed by bacteria and fungi. Decay of the soft parts produces a sticky liquid which glues rock particles together into clumps called *soil crumbs*. Decay of harder tissues, particularly of plants, produces fibres which bind rock particles together. Humus fibres and sticky glue help to prevent soil being blown away by high winds. They also allow rain water to drain freely through the soil, and cause air spaces to form in soil so that plant roots can obtain oxygen. In addition, humus provides plant 'food' in the form of dissolved minerals which are essential for healthy plant growth.

Living organisms

Worms improve soil (make it more fertile) in several ways (Fig. 3). Their burrows allow air to reach plant roots. They add humus to soil by dragging leaves into their burrows and leaving them there to rot. Some worms turn soil over by passing it through their digestive systems and leaving it on the soil surface as worm casts.

Bacteria are extremely important soil organisms. Some bacteria break down dead animals and plants into humus and into essential plant foods called *nitrates*. Other bacteria make nitrates out of nitrogen gas which they obtain from the air.

The best type of soil

The most fertile soil of all is called *loam*. Best loam is 50 per cent sand, 30 per cent clay, 20 per cent humus, and has a thriving population of bacteria and earthworms. Loam is very easy to dig and plough and contains all the dissolved minerals needed for healthy plant growth.

Exercises

1　How are rock particles of soil formed?

2　What is humus, and how is it formed? In what ways do plants benefit from humus?

3　In what ways do worms and bacteria improve soil?

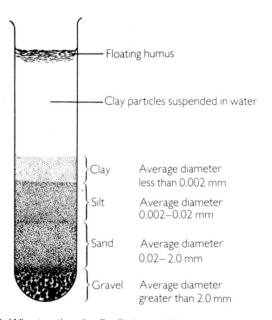

Floating humus

Clay particles suspended in water

Clay — Average diameter less than 0.002 mm

Silt — Average diameter 0.002–0.02 mm

Sand — Average diameter 0.02–2.0 mm

Gravel — Average diameter greater than 2.0 mm

Fig. 1 What is soil made of?   Find out with this simple test. Shake up some soil in water and let it settle. The rock particles settle to the bottom (heaviest first), and humus floats to the top.

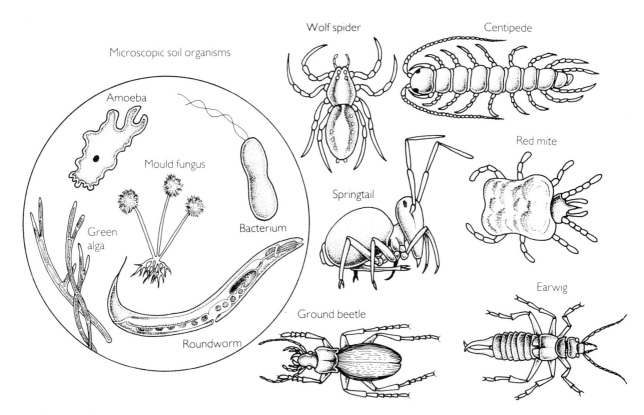

Microscopic soil organisms

Amoeba

Mould fungus

Green alga

Bacterium

Roundworm

Wolf spider

Centipede

Red mite

Springtail

Earwig

Ground beetle

Fig. 2 A few of the many creatures found in soil and among dead leaves

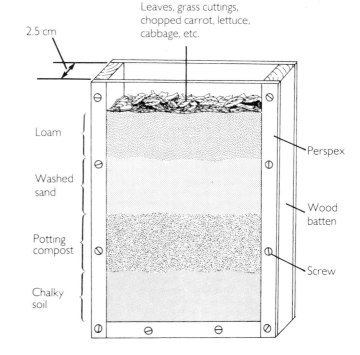

2.5 cm

Leaves, grass cuttings, chopped carrot, lettuce, cabbage, etc.

Loam

Washed sand

Potting compost

Chalky soil

Perspex

Wood batten

Screw

Fig 3 How to study worms Make a wormery about 30 cm square. Fill it with different types of soil, poured in carefully so there is a clear boundary between each layer. Moisten the soil and introduce about ten worms. Scatter some food on the top (see drawing) and put the wormery in the dark. Observe regularly over the next few weeks. What happens to the layers of soil? What happens to the food?

2.5 Insect-eating plants

Reports of jungle plants which trap and eat people are pure fantasy, but there are over 400 types of plants which catch and feed on small animals, especially insects.

Why do some plants eat insects?

Insect-eating plants have chlorophyll and are capable of photosynthesis but they live in waterlogged and acid soils which contain very little nitrate. Nitrate is a chemical rich in nitrogen which plants need to make proteins. Insect-eating plants obtain nitrogen from the bodies of insects which are caught and killed in various ways, and then digested. The Figures in this Unit describe several examples.

Insects and other small animals are attracted to these plants by bright colours, scent, or sugary bait. Then they are trapped, by sticky fluid, by enfolding leaves, or inside various containers. Finally the plant releases digestive juices which dissolve the victim so that its nitrogen-rich chemicals can be absorbed.

Exercises

1   What are the main differences between loam soil and the soils on which insect-eating plants grow?

2   What substance do these plants obtain from the insects they catch? How is this substance extracted from the insects, and what does the plant use the substance for?

3   Describe three different ways in which insects are trapped by these plants.

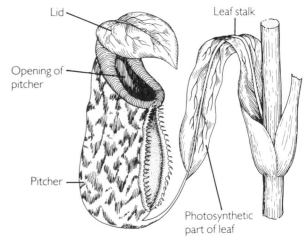

Fig. 1 Pitcher plants have large pitcher-shaped leaves with a hinged lid. Nectar is produced just below the rim of the pitcher. When insects enter to drink it, slippery walls and downward pointing hairs force insects to the bottom of the pitcher where they are drowned and digested in a pool of digestive juices.

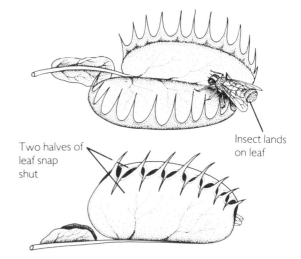

Fig. 2 The Venus fly-trap has leaves with a hinge-like mid-rib, and a row of spikes around the edge. When an insect lands on a leaf the two halves snap together and the spikes interlock so it cannot escape. After the insect has been digested the leaf opens and the undigested remains dry up and blow away.

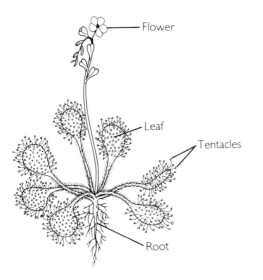

Fig. 3 Sundew lives in marshland among bog mosses. It has leaves with short tentacles covered in sticky glue. An insect landing on a leaf gets stuck, and the tentacles curl over holding it firmly. Glands on the tentacles release digestive juices. Later the tentacles uncurl and the undigested parts of the insect dry up and blow away.

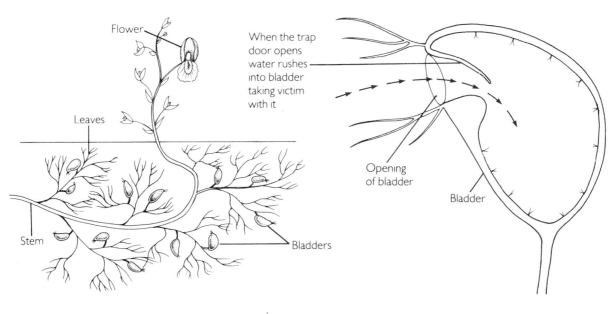

Fig. 4 Bladderwort has no roots. It floats in moorland ponds. It has small bladders each of which has an opening guarded by a trap door. There is less water in the bladder than outside. When a small water animal touches the door it opens and water rushes into the bladder taking the animal with it. The bladder wall produces digestive juices which kill and dissolve the animal.

2.6 Pollution

Pollution damages the living world. The survival of the human race and of all creatures on Earth depends upon our finding ways to stop this damage.

What is a pollutant?

A *pollutant* is a substance which, when it gets into water, air, or soil, harms living things. Pollutants can cause illness, they can prevent organisms reproducing, and they can kill. Pollutants spread through food chains. They are absorbed by plants, then pass to herbivores, then to carnivores, and so on, damaging each link in the chain and sometimes destroying it altogether.

Air pollution

Air is polluted by dust; and by smoke and fumes.

Dust   Dust is made up of solid particles small enough to float for a while in the air. Dust comes from sawing, drilling, and grinding things like rock, metal, wood. Dust also includes the particles of carbon contained in smoke. Dust gathers in the lungs and affects breathing. When it settles on plants it blocks stomata and slows down photosynthesis by reducing the amount of light reaching the leaves.

Smoke and fumes   Every year 1.5 million tonnes of smoke and fumes are discharged over the United Kingdom. Smoke and fumes contain gases such as sulphur dioxide, carbon monoxide, nitrogen oxide, plus lead compounds. Sulphur dioxide and nitrogen oxide gases combine with moisture in the air to form corrosive acids which rot the stonework of buildings, damage plant leaves, and irritate the eyes, breathing passages, and lungs of humans and other animals. Carbon monoxide gas is very poisonous and can kill if inhaled in large amounts. Lead compounds are released in the exhausts of diesel and petrol engines and can cause ill-health.

Water pollution

Water is polluted by sewage from houses and farms, by chemical wastes from factories, and by oil spilled from damaged tankers, oil rigs, etc. Sewage can be made harmless, but in many countries raw sewage is released into rivers and the sea. Sewage is decomposed by bacteria in water, but this uses up oxygen so quickly that fish, insects, tadpoles, etc. can be killed (Fig. 1). Industrial waste often contains very poisonous long-lasting pollutants such as cyanide, lead, mercury, and copper. These chemicals are poisonous even in small quantities. In some modern farms poultry and pigs are kept in buildings and there is no land on which to spread the manure which they produce. The manure is sometimes put into local streams and rivers where it decomposes and reduces the amount of oxygen in the water. Other pollutants released from farms include chemical sprays which kill insect pests, and fungi which attack crop plants. If these chemicals enter rivers and ponds they spread throughout food chains, and can poison human foods.

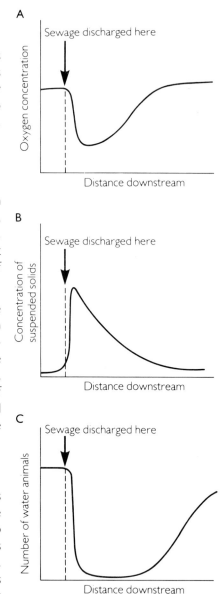

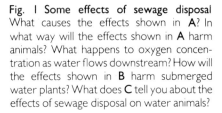

Fig. 1 Some effects of sewage disposal What causes the effects shown in **A**? In what way will the effects shown in **A** harm animals? What happens to oxygen concentration as water flows downstream? How will the effects shown in **B** harm submerged water plants? What does **C** tell you about the effects of sewage disposal on water animals?

Exercises

1 What is a pollutant?

2 Name some substances which pollute air and water, and explain where each substance comes from.

3 In what ways does dust damage plants and humans?

4 What happens to make sulphur dioxide and nitrogen oxide gases particularly dangerous when they are released into moist air?

5 Over the last few years oil companies have reduced by a small amount the lead contained in petrol. Why was this done?

6 Answer the questions asked in the captions to Figures 1 and 2.

Bird killed by oil-polluted water

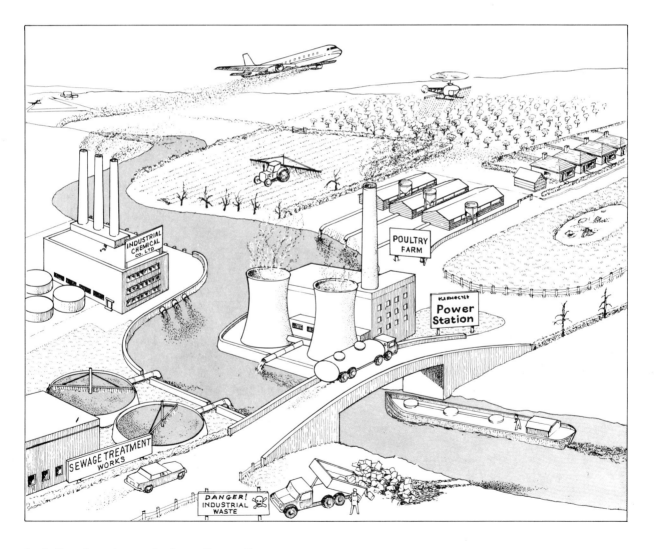

Fig. 2 Describe each type of pollution illustrated here

39

2.7 Conservation

The main aim of nature conservation is to allow people to enjoy the benefits of modern civilization without destroying living things.

Agriculture and conservation

Forests are being uprooted, marshes drained, and moor and heaths ploughed up to make space for the farmland which feeds us all. Once, when wildlife habitats were turned into farmland some plants and animals could live in hedgerows, farm ponds, and small patches of remaining woodland. In modern times this is no longer so easy. Nearly 10 000 kilometres of hedge are uprooted in the United Kingdom each year, and 50 per cent of ponds have been filled in in England since 1945. The remaining hedges and ponds are all too often made uninhabitable by insecticides, fungicides, and weed killers sprayed onto nearby crops.

Food supplies must be maintained and, where possible, increased, but if agricultural improvement continues on its present path future generations may be left with a wildlife consisting of nothing but tough adaptable species like rats, rabbits, and house sparrows.

Some farmers could do far more to protect wildlife. They could decide never to destroy a wood, marsh, hedge, or pond unless absolutely necessary. They could take care not to pollute wildlife habitats, and they could create new habitats by planting native trees and shrubs in parts of their land where farming is impossible.

National parks, reserves, and unspoilt countryside

National parks and nature reserves have been created to preserve nature and for the enjoyment of the public. But even these places, and wild countryside in general, are endangered, often by the people who came to enjoy them.

Some people uproot wild plants, take eggs from nests, set woods and moors alight with cigarette ends and picnic fires, leave fish hooks where they can be swallowed by animals, and pollute an area with discarded rubbish which could have been disposed of properly.

Some teachers and school parties are guilty of destroying what they come to the countryside to study. Flowers need not be picked, they can be identified where they grow. If it is necessary to collect insects and other animals this must be done without trampling all over their habitat. When these animals have been studied it is far better to return them to the wild than to kill and preserve them. Killing animals for a collection should only be attempted as part of a genuine scientific project.

It is now illegal to uproot *any* wild plant without the land owner's permission, and it is illegal even to pick the flowers of certain rare protected species.

Fashion and beauty

Tigers, vicunas, and some species of elephant have been hunted almost to extinction for such things as tiger skin coats, vicuna cloth jackets, and ivory chess sets.

Whales are another example. Several species have been hunted to the point from which they may never recover. Why? Because whale meat is an excellent pet food; because whale oil softens leather used to make gloves; and because millions of women smear their lips with lipstick made from a waxy substance called ambergris found in a whale's stomach. Some perfume manufacturers also use ambergris, together with musk, a substance taken from the rare musk deer. Musk is extracted from a small sac in the deer's abdomen, but the deer must be killed first.

Britain has now banned the use of whale meat as pet food. Glove, lipstick, and perfume manufacturers could use available man-made ingredients which do the same job as whale oil, ambergris, and musk. You can help stamp out trades which exploit endangered species by avoiding their products.

Exercises

1 List the activities illustrated in Figure 1 which people should avoid when visiting the countryside.

2 List examples illustrated in Figure 1 of how the following result in the destruction of wildlife:
   a) the need for more food
   b) the need for better transport
   c) the need for more water
   d) the need for building materials.

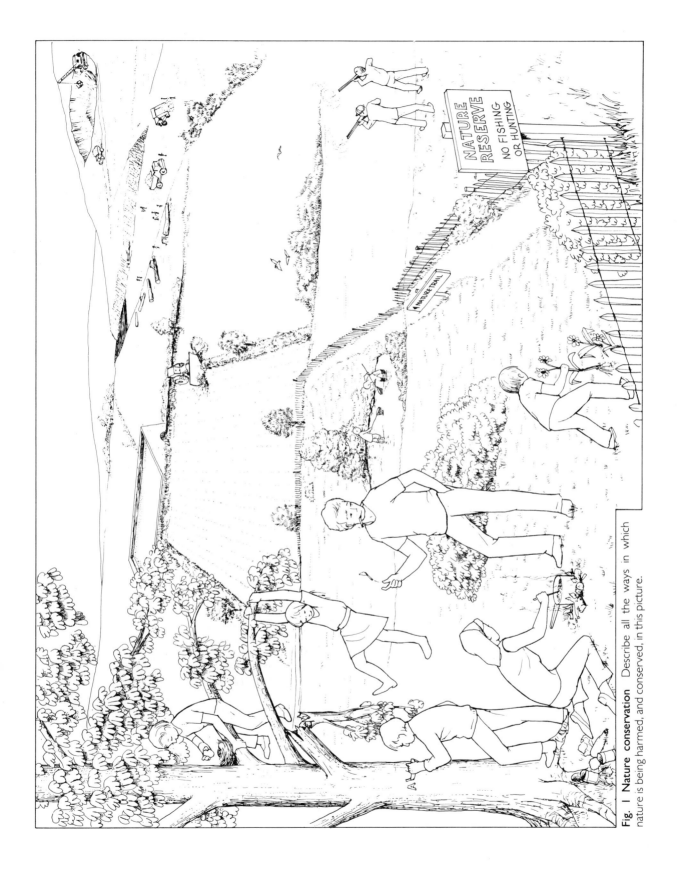

Fig. 1 Nature conservation Describe all the ways in which nature is being harmed, and conserved, in this picture.

2.8 Space biology

We are all space travellers and our spaceship is the Planet Earth.

Spaceship Earth

The Earth is a ball of rock floating in the vacuum of space. We are able to live on it because plants provide oxygen and food and because fungi and bacteria decompose waste and dead bodies, returning them to the soil for plants to absorb. This cycle is known as a *closed ecological system*: nothing is lost or used up because all the materials of life are used over and over again.

Before we can leave Earth and make a long journey into space we must build a spaceship with its own closed ecology. Figure 2 describes how this could be done.

Acceleration and deceleration

During acceleration and deceleration the force of gravity appears to increase. As a modern spaceship accelerates from its landing pad, its astronauts experience a force up to seven times greater than normal gravity (7g). That is, they become seven times heavier. The effects this has on the body are described in Figure 1.

Weightlessness

The weightlessness experienced in space causes several biological problems. The heart has far less work to do because blood too is weightless, and bones and muscles no longer have to support the body so they are under less stress. Long periods in space (one or more years) may weaken the body so much that when astronauts return to Earth the force of gravity may be too much for them.

Exercises

1. a) Explain what is meant by a closed ecological system, using the Planet Earth as an example.
   b) Why would a closed ecology be essential in a spaceship designed for a long voyage, when it was not essential for the voyage to the moon and back?

2. a) What is meant by an acceleration of 7g?
   b) What effects would this have on the body?

3. In what ways could a long period of weightlessness weaken the body?

4. Look at Figure 3. What are the risks from radiation during space travel?

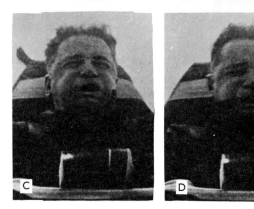
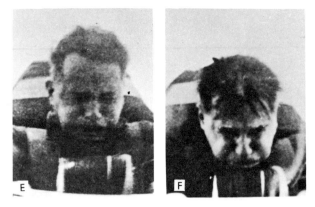

Fig. 1 Effects of acceleration and deceleration The man is strapped to a rocket-powered sled. Photographs **A, B,** and **C** show him experiencing an acceleration force of 12 g. Photographs **D, E,** and **F** show the effects of deceleration forces reaching 22 g. These forces displace organs, make limb movements difficult, blur vision, and slow blood flow, which affects brain activity (thinking). During take-off and landing, astronauts experience these forces for several minutes on end.

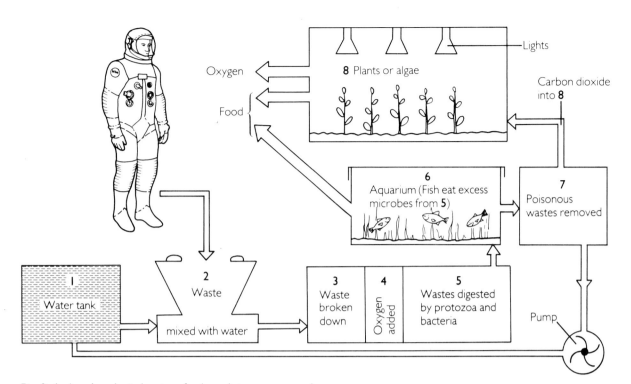

Fig. 2 A closed ecological system for long distance space craft
Plants (**8**) absorb carbon dioxide from respiration, together with
some wastes, and produce oxygen and food. Microscopic algae or
large plants could do this. Sunlight could be reflected onto plants to
save electricity. Fish (**6**) could be replaced by rabbits or chickens,
but these would pose the problem of how to decompose fur,
feathers, and bones, into plant compost. All materials must be used
over and over again with no loss whatsoever.

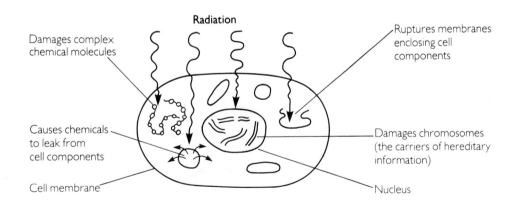

Fig. 3 Effects of space radiation on cells   Air prevents most of
the radiation from space reaching us. To provide equivalent
shielding a space craft would have to have lead walls 33 cm thick.
Dangerous radiation from space includes cosmic rays, and X-rays
and gamma rays from the sun.

Topic 2 exercises

Photosynthesis

1 The bottles in Figure 1 were sealed to make them completely airtight, and were then placed in a well-lit position.

    a) In which bottles will oxygen be produced?

    b) In which bottle will there be the greatest surplus of oxygen?

    c) In which bottle will the snail be able to live the longest: bottle 1 or 2?

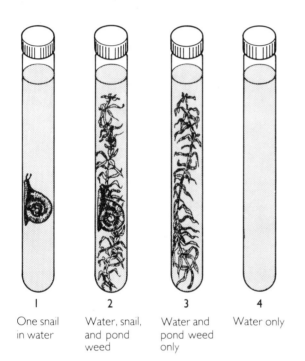

1	2	3	4
One snail in water	Water, snail, and pond weed	Water and pond weed only	Water only

Fig. 1 See exercise 1

Pollution

3 The map in Figure 2 shows the animals which were found in samples of water taken from three different points on a river.

    a) Which animals are found in *all of* clean, partly polluted, and heavily polluted water?

    b) Which animals are found only in clean water?

    c) Which animals seem almost to prefer heavily polluted water?

    d) Which substance, vital to life, will be plentiful in the water at point A, but scarce at point B?

    e) Why is this vital substance likely to be scarce at point B?

    f) What substances are likely to be abundant in the water at point B, but absent from point A?

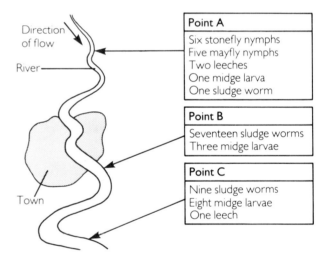

Fig. 2 See exercise 3

Soil

2 Obtain enough soil to fill a plant pot from: a well-tended garden, woodland, meadow, sand dune, river estuary (i.e. as many different types of soil as possible). Place the samples in separate labelled plant pots, then plant seeds in each. Give each pot the same amount of water, light, and heat. In which soil do the seeds sprout first, and in which do they grow best?

Conservation

If you are interested in conservation join *Watch*, the conservation club for schools and young people in general. Details from: *Watch, 22 The Green, Nettleham, Lincoln, LN2 2NR.*

There are many organisations you can join which promote conservation of plants and animals. Details are available in local libraries.

Food and feeding

This rabbit flea is feeding on a rabbit's ear

3.1   Types of food

Our bodies must obtain at least forty different substances from food if we are to remain alive and healthy.

Why do we need food?

Food provides energy to work the muscles and organs of the body, and the materials needed for growth and repair of damaged and worn-out tissues. The energy the body obtains from food is now measured in kilojoules (abbreviated to kJ). Calories are no longer used in scientific work (1 Calorie = 4.2 kJ).

Foods can be arranged into the following groups: carbohydrates, proteins, fats and oils, vitamins, and minerals.

Carbohydrates

Carbohydrates are sugary and starchy foods. These are the body's main source of energy (1 gram of carbohydrate provides 17 kJ of energy). Sugar is all carbohydrate, while jams, potatoes, rice, bread, cakes, buns, and fruit are very rich in this food. Milk, peas, and beans have some carbohydrate.

Proteins

The body uses proteins for growth, and for repairing damaged and worn-out tissues. In other words, proteins are body-building foods. Proteins are not usually used for energy but can provide 17 kJ of energy per gram of protein. Large amounts of protein are found in meat, fish, cheese, and eggs. Milk, bread, peas, and beans contain some protein.

Fats and oils

These are the body's main stored foods. When they are eventually used in respiration 1 gram of fat provides 39 kJ of energy. A layer of fat under the skin helps reduce the rate at which the body loses heat in cold weather. Butter and lard are all fat. Meat is often surrounded by fat. Milk, nuts, and certain fish contain some oil, and fat is a basic ingredient in cake and pastry making.

Vitamins

Vitamins are essential for growth and general health even though we require them in extremely small amounts—often less than one-thousandth of a gram a day. Table 1 gives details of the most important vitamins.

Minerals

We require about fifteen different minerals to remain healthy. Most are found in meat, eggs, milk, green vegetables, and fruit. Common salt enables messages to travel along nerves. Muscles need potassium and sodium to contract. Teeth and bones are mostly calcium, magnesium, and phosphorus. Red blood cells contain iron.

Vitamin	Best food source	Functions in body
A	Fish liver oils, liver, green vegetables, milk, carrots	Keeps skin, bones, and eyes healthy. Helps prevent nose and throat infections.
B (Group of vitamins)	Yeast, wholemeal bread, green vegetables, eggs, liver, milk, cheese, meat, fish	Growth, release of energy from food, health of blood, eyes, and skin. Lack of B vitamins causes beri-beri (paralysis of limbs) and pernicious anaemia (poorly formed red blood cells).
C	Oranges and lemons, blackcurrants, green vegetables, tomatoes, potatoes	Helps wounds to heal. Health of gums and teeth. Helps prevent nose and throat infections. Lack of vitamin C causes scurvy (soft gums, loose teeth, and poor healing of wounds).
D	Liver, butter, cheese, eggs, fish. Made by skin when exposed to sun	Needed to make bones and teeth. Lack of vitamin D causes rickets (soft and weak bones).

Table 1   Vitamins

Exercises

1 Which of the following statements describe carbohydrates? proteins? fats and oils?
   a) Body-building foods.
   b) The body's main stored foods.
   c) Sugary and starchy foods.
   d) Provide materials for growth and repair.
   e) Provide 39 kJ of energy per gram.
   f) The body's main source of energy.

2 Which of the following statements describes common salt? potassium? calcium? and iron?
   a) Found in bones.
   b) Found in red blood cells.
   c) Needed by muscles to contract.
   d) Enables messages to travel along nerves.

3 Study the samples of foods illustrated in Figure 1.
   a) Which contain large amounts of carbohydrate? of protein? and of fats or oils?
   b) Which are good sources of vitamins? and minerals?

4 Which vitamin, or vitamins, are described by each of the following statements?
   a) Found in liver.
   b) Found in oranges and lemons.
   c) Found in eggs.
   d) Help prevent nose and throat infections.
   e) Helps wounds to heal.
   f) Needed to make bones and teeth.
   g) Release energy from food.
   h) Made in skin exposed to sunlight.

Fig. 1 **Some examples of food** (see exercises)

3.2 Diet and health

If you get your diet right your body will run as well as it can.

A balanced diet

A balanced diet is one in which energy-giving foods (carbohydrates and fats) in each meal are balanced by body-building foods (proteins) and protective foods (vitamins and minerals). A balanced meal is made up of at least one part protein, no more than one part fat, and four parts carbohydrate, and must include foods such as salads which are rich in vitamins and minerals.

Food eaten must balance energy used

It is very important that the amount of food eaten each day should supply no more and no less than the energy used in that day. If you regularly eat an amount of food which supplies more energy than you need, the surplus food will be converted into fat and this could make you overweight (Fig. 1).

Obesity

Those who are overweight (obese) risk developing diseases of the heart and blood vessels in which arteries become clogged with fatty material called cholesterol; and diseases of the liver, kidneys, and gall bladder. They are more likely than slim people to have high blood pressure, gout, diabetes, digestive disorders, and varicose veins. The more you are overweight, the more you endanger your life.

Slimming—dodge the stodge

To reduce weight or stop becoming overweight avoid sugar, pastry, cakes, biscuits, and fried and fatty foods. Why? Because small amounts of these foods supply huge amounts of energy. Consequently, you eat enough to satisfy your daily energy needs long before you satisfy your appetite.

Compare these figures. Five chocolate biscuits (about 100 g) supply enough energy for an average-sized man to cycle for 87 minutes at 15 kph. But the *same weight* of wholemeal bread supplies only enough energy for 36 minutes' cycling at this speed.

This means you don't have to avoid *all* carbohydrates to remain slim. Just *avoid* the stodgy, energy-rich ones like sugar and sweets, biscuits, cakes, puddings, pastry, tinned fruit in syrup, cream, butter, chips, and crisps. Instead, your daily ration of carbohydrate should come from *moderate amounts* of wholemeal bread, rice, potatoes, steamed fish, grilled meats, and cheese. Don't eat more than 50 grams of fat a day. But remember that this is total fat: don't forget the fat hidden away in pastry, cheese, meat, and milk. Eat *as much as you like* of fresh fruit, salads, green and root vegetables, kidney, heart, cottage cheese, yoghurt, and thin soups.

Exercises

1 What is meant by the phrase 'a balanced diet'?

2 Why is it important for one day's food to supply no more than one day's energy?

3 Use Table 2 to find out how much energy you use in a day. Make a list of the food you eat in one day and estimate its weight. How does it compare with the recommended weights given in this table?

4 Why is it dangerous to be overweight?

5 List some foods which should be avoided by people who wish to lose weight. Which foods can they eat any amount of?

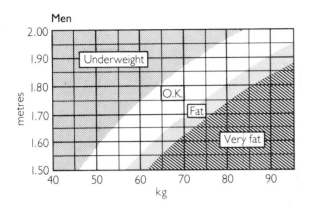

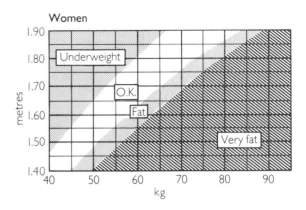

Fig. I About half the population of Britain is overweight

Age, sex, occupation	Energy used in I day		Diet (daily)
15 years 	**Males** 12 600 kJ	**Females** 9600 kJ	
Adult (light work)	11 550 kJ	9450 kJ	80–100 g of protein a day; 300 g of carbohydrate (except for those doing heavy work who should eat far more), and 50–100 g of fat. Large amounts of sugar should be avoided as sugar increases tooth decay.
Adult (moderate work)	12 100 kJ	10 500 kJ	
Adult (heavy work)	15 000 kJ to 20 000 kJ	12 600 kJ	

Table 2 The amount of energy used each day

3.3 Feeding a baby

The best baby food is free, immediately available, already sterilized, and perfectly balanced.

Breast milk is best

Both bottle feeding and breast feeding can produce babies which are mentally and physically healthy. However, there are certain differences between these two feeding methods which can give breast-fed babies many advantages.

Breast milk is perfectly suited to babies Bottle milk is made from dried, processed cow's milk with added artificial ingredients. The processing destroys many important food substances in the milk. Breast milk is pure and fresh, and its contents change to meet exactly the needs of a growing baby.

Breast milk helps fight germs Breast milk contains substances called antibodies. These destroy many germs that threaten babies. Breast milk also contains living cells which act against germs in an infant's stomach. Because of this, breast-fed babies have less diarrhoea, less vomiting, fewer skin disorders such as nappy rash, and they probably have fewer lung infections, such as bronchitis and pneumonia, than bottle-fed babies.

Breast milk protects against allergies When compared with bottle-fed babies, those fed on breast milk are less likely to have allergies such as sensitivity to substances like pollen. So they are less likely to have hay fever and similar disorders.

Breast milk aids brain development Breast milk contains large amounts of taurine, a substance which is important for the development of the central nervous system. Very little taurine is present in bottle milk.

Breast milk is digested more quickly than bottle milk This is one reason why breast-fed babies rarely suffer from constipation and the painfully hard bowel movements which frequently upset bottle-fed babies. Consequently breast-fed babies often cry less.

Breast milk helps prevent tooth decay Research has shown that breast-fed babies have fewer bad teeth than bottle-fed babies. And sucking at the breast helps teeth to become correctly positioned in the jaw bones.

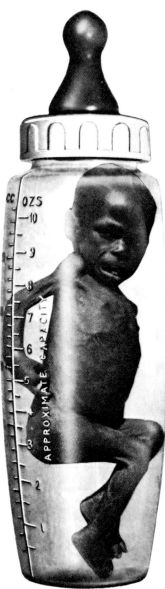

Fig. 1 Many mothers in poor countries have no clean water supplies, limited cooking facilities to sterilize baby bottles, and are often unable to read the instructions on powdered milk containers (even if written in their own language). Yet many companies encourage mothers to give up breast feeding and change to powdered milk. The result has been an increase in child starvation, illness, and death, because these mothers use dirty water in dirty bottles, and over-dilute the milk because of its high cost.

| | | Breast | Cow | Commercial milks | | | |
				Brand A	Brand B	Brand C	Soya
Protein	g	1.4	3.4	1.45	1.5	1.5	2.2
Fat	g	4.0	3.9	3.8	3.8	3.6	3.2
Lactose sugar	g	7.0	4.6	7.0	7.2	7.2	7.3
Energy	kJ	289	280	284	284	275	273
Minerals:							
Calcium	mg	34	124	36	40	44	59
Sodium	mg	15	52	19	18	15	32
Potassium	mg	57.5	155	57	60	56	68
Chloride	mg	43	98	44	57	40	43
Magnesium	mg	3.4	12	5.2	4.5	5.3	5
Phosphorus	mg	15	98	31	27	33	43
Vitamins:							
A	µg	56.5	40	100	80	79	85
D	µg	0.01	0.02	1.0	1.1	1.1	1.2
E	mg	0.45	0.09	0.48	1.0	1.0	1.6
K	µg	1.7	–	2.7	2.8	5.8	2.7
B_1	µg	16	40	42	70	80	95
B_2	µg	36.8	200	55	100	110	135
niacin	µg	201	80	690	850	1000	946
C	mg	4.1	1.5	6.9	5.5	5.8	6.8

Table 3 Composition per 100 ml of human breast milk, cow's milk, and four commercial bottle-feeding milks.

Bottle feeding

Sometimes a mother cannot breast-feed her baby, and has to rely on a substitute milk. The obvious choice is cows' milk. But untreated cows' milk is not fully absorbed by a baby's body and often causes indigestion. So before cows' milk is fed to human babies, it is modified. Changes are made in the composition of the milk to make it match human breast milk as closely as possible.

If you look at the table you will see that cows' milk contains twice as much protein as human milk. This is because a calf needs to grow twice as fast as a baby. So the quantity has to be adjusted to 'baby-level'. This has been done with the three commercial milks shown in the table (Brands A, B, and C).

Another difference between cows' milk and human milk is in the amount of lactose sugar present. The amount has been nearly doubled in the commercial milks in order to match the amount in breast milk.

Vitamin and mineral levels need adjusting before cows' milk is suitable for babies. You can see from the table that a calf needs much higher levels of minerals than a baby. On the other hand, a baby needs more of some vitamins, like vitamin C, and less of others, like vitamin B, than a calf.

Sometimes a baby is allergic to cows' milk. Some companies make a baby-milk from soya beans which can be fed to such babies. Look at the last column in the table. The soya milk does not match breast milk as closely as the cows' milk products, but it is an alternative if the baby cannot be fed other commercial milks.

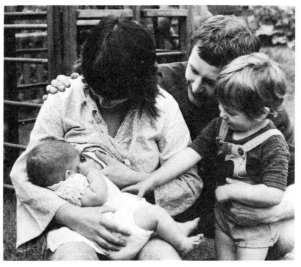

Human milk is perfect for human babies. Breast-feeding advice and information can be obtained from La Leche League of Great Britain, BM 3424, London WC1V 6XX.

3.4 Wholefoods and processed foods

Processed foods are altered in many ways before you buy them. These alterations usually remove something useful, and often add something which may be harmful.

The differences between wholefoods and processed foods

Wholefoods are as close to natural as possible—they are changed little if at all before they are eaten. Unlike processed foods they are free from chemical preservatives, and from artificial colourings and flavouring. The best wholefoods are nuts, fresh vegetables and fruits, meat, fish and eggs bought unprocessed and cooked at home, honey, full cream milk, and natural yoghurt.

Food is processed to preserve it, and also to reduce the time needed to prepare it for eating. These seem good ideas if you have little time to spare for shopping and cooking. But remember, the way your body performs, feels, and looks, depends to a large extent upon how you feed it, so it makes sense to feed it with the best food you can get.

What processing removes from food

Roughage   This is plant fibre and is found in vegetables, fruit, and especially in the bran which surrounds wheat grains (Fig. 1A). Roughage keeps the digestive system in good working order. It soaks up water and poisonous wastes from food during digestion and becomes a firm bulky mass which muscles in the gut wall can grip and move easily along the digestive passageways and out of the body when you use the lavatory.

Food without roughage forms hard dry lumps of waste which get stuck in the gut causing constipation. Sometimes the lumps of waste rot, causing swellings which require surgery. People whose diet does not contain enough roughage are more likely to suffer from bowel cancer.

Bran is removed during the milling of white flour (Fig. 2), and fibre is removed when vegetables and fruit are canned or dried.

Vitamins and minerals   Both of these are lost when wheat germ (Fig. 1A) is removed from wheat grains during the milling of white flour and in the manufacture of some breakfast cereals (Fig. 2). Food is always boiled before it is canned, dried, or frozen, and before drying it is chopped up and exposed to air. Both these processes destroy vitamin C and the boiling dissolves away minerals from food. During processing many foods are treated with sulphur dioxide gas to preserve them. This gas destroys vitamin B_1. Finally, more vitamins and minerals are lost when cans and packets are opened and their contents heated up before a meal.

What processing adds to food

Some common food additives are described on the next page.

A   Parts of a wheat grain

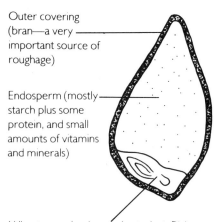

Outer covering (bran—a very important source of roughage)

Endosperm (mostly starch plus some protein, and small amounts of vitamins and minerals)

Wheat germ (embryo wheat plant. Rich source of fatty acids, vitamin B, iron, magnesium, calcium, copper, phosphorus, potassium, manganese)

Flour milling

B   All of the wheat grain is used to make wholemeal bread

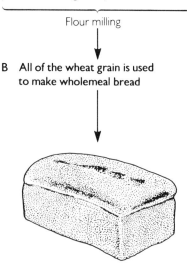

Fig. 1 Don't say brown, say wholemeal
No part of the wheat grain is removed from flour used to make wholemeal bread. Do not confuse it with brown bread which is only white bread plus colouring.

Antioxidants These stop fats going rancid. Their use is restricted by law because they are potentially dangerous to health.

Colourings These restore colour lost in processing, and make food appear to have more of an ingredient than it really does.

Emulsifiers These allow water and oil to mix.

Flavourings These restore flavour lost during processing, and make food appear to have an ingredient it doesn't: 'strawberry flavour' means a food has a strawberry flavoured chemical and not necessarily real strawberries.

Preservatives These slow the rate at which food goes bad. Common preservatives are sulphur dioxide, nitrites and nitrates, and benzoic acid, all of which are potentially dangerous.

Stabilizers These preserve the texture of food, and help prevent emulsions separating.

Exercises

1 Visit a supermarket and read the labels on processed food. Note which foods have the additives listed above. Make a chart with additives across the top and foods down the left side. Fill in the chart with ticks to indicate which food has which additive.

2 Devise meals using the foods in the chart. How many additives would you consume if you ate these meals?

3 Devise meals avoiding as many additives as possible.

4 Visit a variety of food shops—small grocers, health food shops, greengrocers, etc. How many foods can you find without any additives? Compare the cost of the same food when it is fresh and when it has been processed.

Inside a modern flour mill

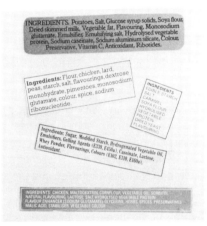

These labels show that sometimes packaged food contains more chemicals than actual food.

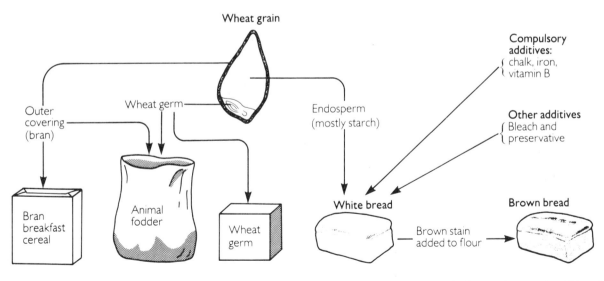

Fig. 2 White and brown bread are made from flour which has had bran and wheat germ removed from it during milling. The flour is then bleached to make white bread and stained to make brown bread. Some vitamins are added but not in amounts which match wholemeal flour. Lost minerals are not replaced.

3.5 Feeding together: commensalism and symbiosis

Sometimes two entirely different organisms live together and form a kind of partnership. If only one of them benefits the partnership is called commensalism, but if both benefit it is called symbiosis.

Commensalism

Sharks, pilot fish, and suckerfish   Sharks often have suckerfish attached to them and pilot fish swimming nearby (Fig. 3). The shark does not benefit from this partnership but the sucker fish hitches a free ride and shares the shark's discarded food with the pilot fish. Both the smaller fish benefit from the fact that their large companion scares away, or eats, animals which would otherwise attack them.

Symbiosis

Hermit crabs and sea anemones   A hermit crab has a long curved abdomen which it pushes into an empty whelk shell for protection. It gets extra protection by finding a sea anemone and, with its pincers, placing it on top of its shell (Fig. 1). The crab is protected from attackers by the anemone's stinging tentacles, and the anemone benefits by eating scraps of food discarded by the crab.

Symbiotic cleaners   Cattle egrets (Fig. 2) and other birds spend their lives eating ticks and parasitic insects from the backs of cattle, antelopes, and rhinoceroses. There is an Egyptian plover which walks boldly into the open mouth of a crocodile to peck leeches from between its teeth, and in the sea over forty different fish get food by cleaning parasites off the gills and bodies of other fish.

Symbiotic digesters   Rabbits, sheep, cattle, and other herbivores eat grass but cannot digest it by themselves. This job is done for them by millions of bacteria which live in their digestive systems. The bacteria obtain shelter, protection, and food and the herbivore has its food changed into a form which its body can absorb.

Insects and flowers   When bees and other insects visit flowers to sip nectar they come away with pollen stuck to their bodies, and transfer it to the next flower they visit. In this way insects get food and plants are cross-pollinated (Fig. 6).

Exercises

Answer the questions in the captions of Figures 1 to 6.

Fig. 1 Hermit crab and sea anemone   What benefits do the crab and the anemone gain from this partnership?

Fig. 2 Cattle egrets and cattle   Cattle have poor vision and hearing, have parasites attached to their skins, and stir up insects with their feet. Egrets have excellent vision and hearing, and eat insects. How do these animals help one another?

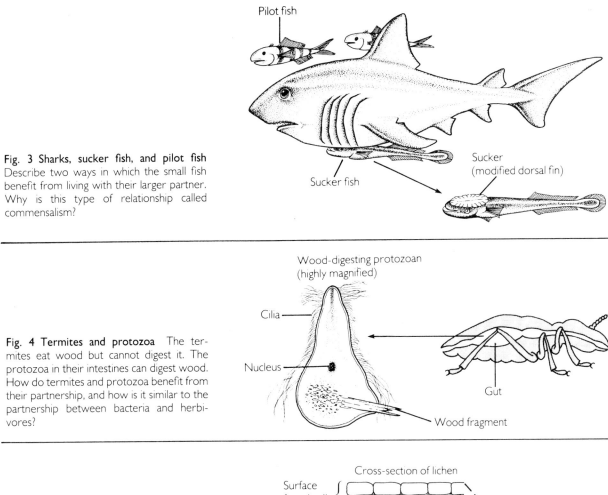

Fig. 3 Sharks, sucker fish, and pilot fish
Describe two ways in which the small fish benefit from living with their larger partner. Why is this type of relationship called commensalism?

Pilot fish

Sucker fish

Sucker (modified dorsal fin)

Fig. 4 Termites and protozoa The termites eat wood but cannot digest it. The protozoa in their intestines can digest wood. How do termites and protozoa benefit from their partnership, and how is it similar to the partnership between bacteria and herbivores?

Wood-digesting protozoan (highly magnified)

Cilia

Nucleus

Wood fragment

Gut

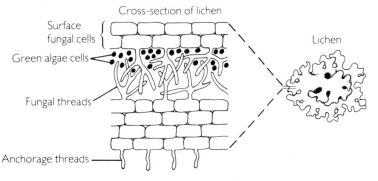

Fig. 5 Lichens are a partnership between green algae and a fungus The algae release food and oxygen during photosynthesis, and the fungus encloses the algae and absorbs water and minerals. What benefits do each gain from the partnership? Is it commensalism or symbiosis?

Cross-section of lichen

Surface fungal cells

Green algae cells

Fungal threads

Anchorage threads

Lichen

Fig. 6 Flowers and nectar-seeking insects What attracts bees to flowers? How does a flowering plant benefit from attracting bees?

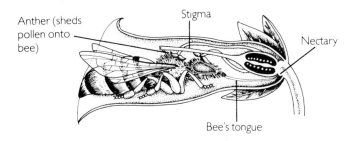

Anther (sheds pollen onto bee)

Stigma

Nectary

Bee's tongue

3.6 Feeding off others: parasites

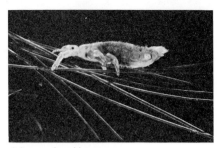

A human head louse

A parasite gets its food from the living body of another organism called the host. A parasite nibbles pieces off its host, sucks the host's blood, etc., but unlike a predator, a parasite does not deliberately try to kill its host as this would kill its source of food.

How parasites live

Feeding Mosquitoes, fleas, and lice have a mouth which is a fine hollow tube, like a hypodermic needle. This tube is pushed into the host's skin until it reaches a blood vessel from which blood is sucked. Leeches and hookworms have teeth which tear through skin making it bleed. Most blood suckers inject a chemical into their host's blood which delays its clotting and blocking up the wound. This means they can take their time having a good meal.

Tapeworms live in their host's intestine. They don't have to digest their food because the host does it for them. In fact tapeworms don't even have a gut, they simply absorb the host's digested food directly through their skin.

Figure 2 shows some parasites of humans.

Finding a host A mosquito flies from host to host, a flea jumps, a bed bug and lice walk, and a leech crawls. A tapeworm has no legs or sense organs so it can hardly move at all and would die in a few minutes if it left its host. Tapeworms spread from host to host when they repro-duce. A tapeworm lays many eggs—beef tapeworms lay 600 million a year. Eggs pass out of the host in its faeces and if sewage disposal arrangements are poor, the eggs may be deposited where cattle eat and drink. If a cow eats a beef tapeworm egg it develops into a larva embedded in the cow's muscle. If humans eat this muscle (beef meat) partly cooked, the larva quickly becomes an adult tapeworm in the human's gut.

Exercises

1   Normal plants have green leaves with chlorophyll to make food by photosynthesis, roots which absorb water and minerals from soil, and seeds which sprout early in spring to give them a long growing season. How is Dodder (Fig. 1) different from this description? Account for the differences.

2   Most animals have limbs, fins, or wings, sense organs, an intestine, and produce fairly small numbers of eggs. How is a tapeworm different from this description? Account for the differences.

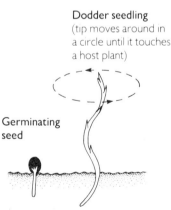

Dodder seedling
(tip moves around in a circle until it touches a host plant)

Germinating seed

Fig. 1 Dodder is a parasite of nettles It has colourless scale leaves without chlorophyll and suckers which suck food from the nettle. Dodder seeds sprout in mid-summer, by which time nettles are fully developed. The tip of the seedling moves round in a circle. If it touches a nettle it twists rapidly up its stem. The Dodder root then dies.

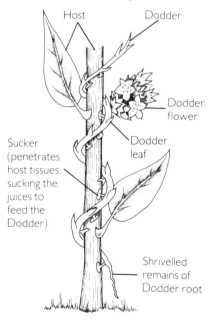

Dodder attached to host plant

Host   Dodder

Dodder flower

Sucker (penetrates host tissues, sucking the juices to feed the Dodder)

Dodder leaf

Shrivelled remains of Dodder root

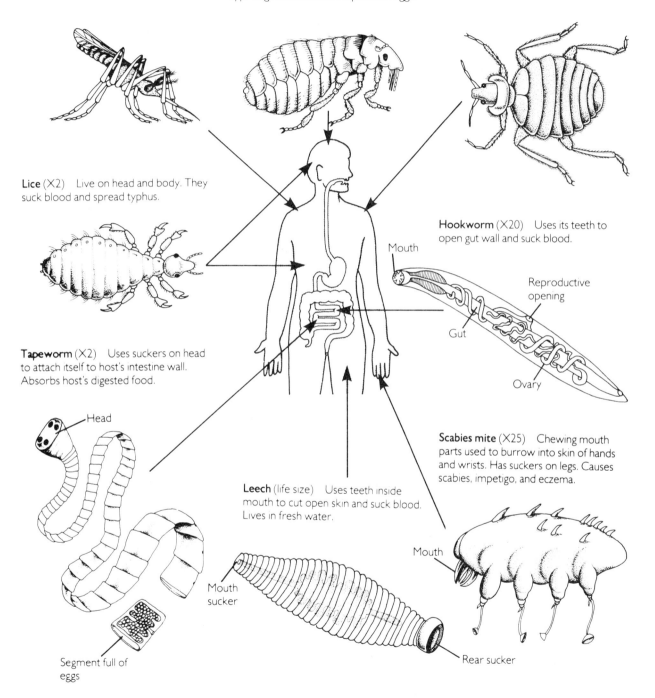

Mosquito (×4) Females suck blood and spread malaria and yellow fever. They use their antennae to detect heat given off from a victim's body.

Flea (×10) Body flattened from side to side, which enables it to move easily between victim's hairs. Has long claws to cling to host. Sucks blood and spreads typhus germs and some tapeworm eggs.

Bed bug (×15) Sucks blood. Bite irritates victim's skin. No evidence that it transmits disease.

Lice (×2) Live on head and body. They suck blood and spread typhus.

Hookworm (×20) Uses its teeth to open gut wall and suck blood.

Mouth

Reproductive opening

Gut

Ovary

Tapeworm (×2) Uses suckers on head to attach itself to host's intestine wall. Absorbs host's digested food.

Head

Scabies mite (×25) Chewing mouth parts used to burrow into skin of hands and wrists. Has suckers on legs. Causes scabies, impetigo, and eczema.

Leech (life size) Uses teeth inside mouth to cut open skin and suck blood. Lives in fresh water.

Mouth

Mouth sucker

Segment full of eggs

Rear sucker

Fig. 2 Some parasites of humans

3.7 Teeth

Teeth are covered with enamel, the hardest substance in the body. But if you eat sweets and fail to clean your teeth properly, it is likely that decay will have made you toothless by middle age.

Teeth and chewing

Teeth grow in holes in the jaw bones called *sockets*. Figure 1 illustrates the four different types of teeth and Figure 2 illustrates the parts of a tooth. Chew food thoroughly before swallowing as this mixes it with saliva and breaks it into small fragments which are quickly digested.

What causes tooth decay?

Tooth decay is caused by *plaque*—a sticky film of food, saliva, and bacteria which forms on teeth after meals. If you eat sweet foods between meals plaque on your teeth absorbs the sugar like a sponge. Bacteria in the plaque then transform this sugar into acid, which dissolves away tooth enamel, eventually making a hole.

But this is not the whole story. Plaque builds up where the teeth and gums meet and can cause a space to form between the gum and a tooth (Fig. 3). Bacteria in this space damage fibres holding the teeth in place, causing the teeth to become loose and fall out. This type of gum disease is the main cause of tooth loss in middle age.

Make your teeth last a lifetime

Two things will help you keep your teeth: *avoid sugar* between meals, and *brush your teeth regularly* using a fluoride toothpaste.

Sugar If plaque is starved of sugar it produces very little acid, and most of this is neutralized by saliva so hardly any damage is done to the teeth. In addition the surface of a tooth can actually 'heal' by forming new enamel. Healing occurs much faster if fluoride is present in the mouth, so always use a fluoride toothpaste.

Brushing teeth Brush your teeth after breakfast and last thing at night, as explained in Figure 4. It takes three minutes to do this properly. It is especially important to remove plaque from the edges of the gums in order to prevent gum disease.

Exercises

1 How many incisors, canines, premolars, and molars do you have? How is the shape of an incisor and a molar related to its function?

2 What is plaque? How does plaque damage tooth enamel, and cause teeth to fall out?

3 If you must eat between meals, why is it better to eat nuts and fruit rather than sweets, biscuits, and chocolate?

4 Why is it important to use fluoride toothpaste?

5 What is the hardest substance in the body?

6 Describe the structure of a tooth. Where is dentine found in a tooth?

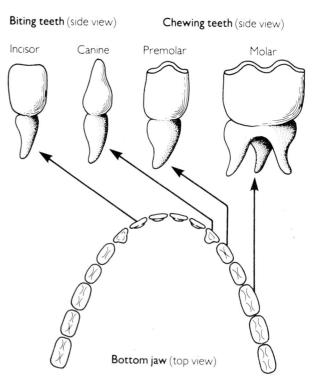

Biting teeth (side view)   **Chewing teeth** (side view)

Incisor   Canine   Premolar   Molar

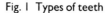

Bottom jaw (top view)

Fig. 1 Types of teeth

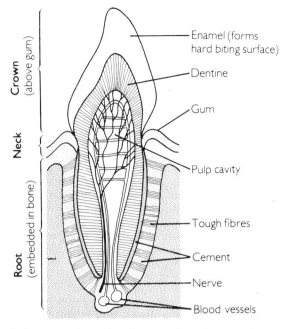

Crown (above gum)

Neck

Root (embedded in bone)

Enamel (forms hard biting surface)

Dentine

Gum

Pulp cavity

Tough fibres

Cement

Nerve

Blood vessels

Fig. 2 Structure of a tooth The root of a tooth is attached to the jaw bone by cement and tough fibres. The crown is the biting surface of a tooth and is made of enamel. The inner part of a tooth is made of bone-like material called dentine, and contains a pulp cavity filled with nerves and blood vessels.

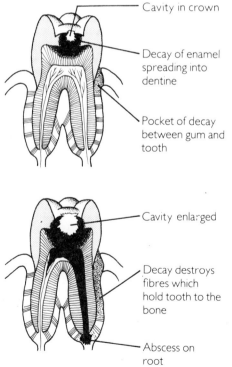

Cavity in crown

Decay of enamel spreading into dentine

Pocket of decay between gum and tooth

Cavity enlarged

Decay destroys fibres which hold tooth to the bone

Abscess on root

Fig. 3 Decay of a molar tooth

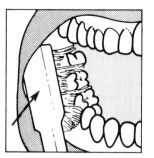
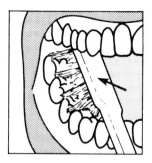

Brush the bottom teeth upwards (back and front) and the top teeth downwards.

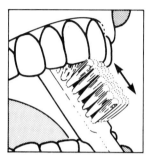
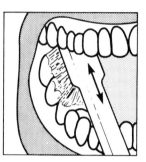

An up-and-down action cleans behind the front teeth and a back-and-forth action cleans the tops of the molars.

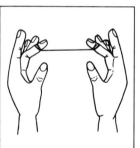
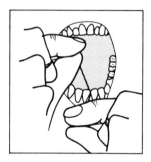

Use dental floss to clean between the teeth. Take about 50 cm of floss thread, wind one end around the middle finger of each hand and, holding your hands about 3 cm apart, pull the floss up and down between the teeth to clean plaque off the side surfaces. Be careful not to cut the gum.

Fig. 4 Care of teeth Clean teeth after breakfast and last thing at night.

3.8 Digestion

Your body cannot use food in the form in which you eat it. This is because bits of chewed-up bread, meat, cabbage, etc., would clog up the blood vessels which transport food around the body, and would be far too big to enter body cells.

Digestion

Digestion makes food soluble so it can pass into the blood as a solution, and then pass from the blood into body cells. Digestion takes place in a tube called the gut, or *alimentary canal*, which starts at the mouth and ends at the anus (Figs. 1 and 2). In certain places the gut walls produce digestive juices with chemicals called digestive *enzymes*. These enzymes break down food into its soluble components (Fig. 3).

Absorption

Absorption is the movement of digested (soluble) food through the gut wall into the bloodstream. Food is transported by blood to the liver where it is processed in various ways and released again into the blood as fast as the body needs it. Blood transports this food to the cells where it is *assimilated*.

Assimilation

This is the movement of food into cells and its use there to provide energy, and for growth and repair.

Elimination (defecation)

Most foods contain substances which cannot be digested. These include cellulose walls of plant cells and plant fibres. Indigestible substances are called *faeces*, and are passed out of the body through the anus. This process is called *elimination*, or *defecation*.

Exercises

1   What is digestion, and why must food be digested before the body can use it?

2   What do digestive enzymes do to food?

3   What is absorption? What happens to food after it is absorbed?

4   What is the scientific name for the tube where digestion takes place? What is a common name for this tube?

5   List four foods which contain substances that cannot be digested. Why are these foods important in the diet? (p. 52 will help you)

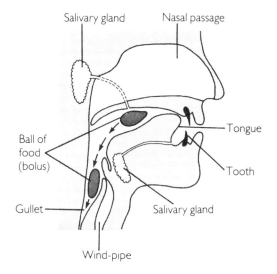

Fig. 1 Digestion in the mouth and swallowing   Chewing mixes food with saliva from the salivary glands. Saliva moistens food and contains an enzyme which breaks down cooked starch into maltose sugar. During swallowing the tongue pushes food to the back of the mouth from where it passes into the gullet. The soft palate prevents food from entering the nasal cavity and the entrance to the wind-pipe is also closed.

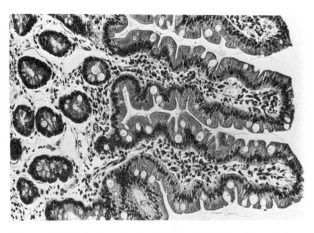

Digested food is absorbed in the small intestine through these finger-like projections. They are called villi and they are shown here magnified 400 times.

60

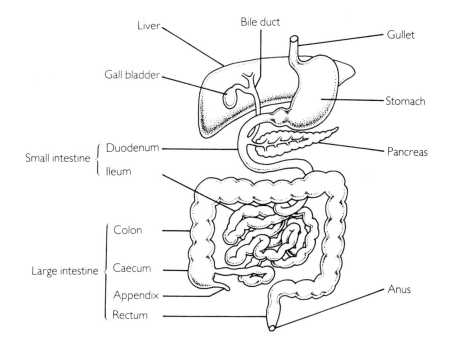

Fig. 2 The gut partly unravelled

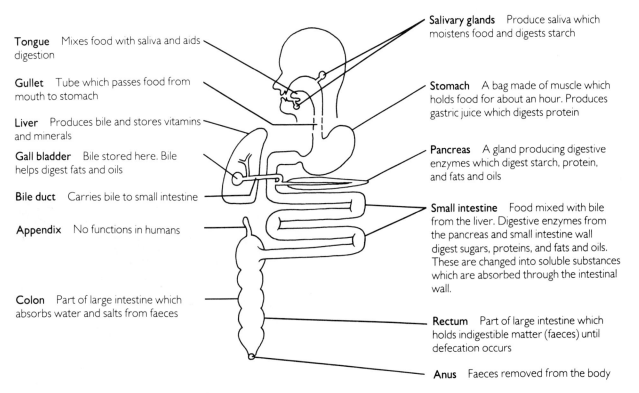

Tongue Mixes food with saliva and aids digestion

Gullet Tube which passes food from mouth to stomach

Liver Produces bile and stores vitamins and minerals

Gall bladder Bile stored here. Bile helps digest fats and oils

Bile duct Carries bile to small intestine

Appendix No functions in humans

Colon Part of large intestine which absorbs water and salts from faeces

Salivary glands Produce saliva which moistens food and digests starch

Stomach A bag made of muscle which holds food for about an hour. Produces gastric juice which digests protein

Pancreas A gland producing digestive enzymes which digest starch, protein, and fats and oils

Small intestine Food mixed with bile from the liver. Digestive enzymes from the pancreas and small intestine wall digest sugars, proteins, and fats and oils. These are changed into soluble substances which are absorbed through the intestinal wall.

Rectum Part of large intestine which holds indigestible matter (faeces) until defecation occurs

Anus Faeces removed from the body

Fig. 3 Diagram of the gut and functions of its parts

3.9 Pests and food production

Farmers have the job of feeding the world's four thousand million mouths. Their job is made harder by pests which attack crops and animals, destroying millions of tonnes of food a year.

Pest control

Effective pest control depends upon a knowledge of how the weather affects pests, upon the use of chemical sprays, and upon breeding plants and animals which can resist pests.

Weather A knowledge of how the weather affects pests helps farmers to know when they are most likely to attack. Slug numbers increase during wet weather, greenfly increases when it is dry, red spider increases when it is hot, and fungal diseases are likely to spread quickly in warm, damp (humid) conditions. Locust swarms, which in a single day can eat enough wheat to feed five million people, often occur after unusually heavy rains.

Chemical sprays Insecticide and fungicide sprays are our strongest weapons against pests, but the odds are still with the bugs, for two reasons. First, they breed at such a fast rate that if you kill 99 per cent of them they are back again in the same numbers after a surprisingly short time. Second, bugs can become resistant to chemical sprays. In a population of millions there will be a few bugs which are not much affected by a particular spray. While their companions die they continue breeding and build up a new population against which the spray is useless.

Fortunately, humans can play the same game as the bugs; by breeding crops and animals which can resist *them*.

Exercises

1 Study the Figures in this Unit. Make a list of the pests which harm the crops (fruit, vegetables, etc.) directly, and those which harm the parent plant. Which of these pests will increase during wet weather? hot weather? humid weather? and dry weather?

2 A farmer's stable was infested with flies so he sprayed it with insecticide. Nearly all the flies were killed. Two weeks later the number of flies was the same as before so he sprayed the stable again. Most of the flies were killed. The flies increased in numbers again and the stable was sprayed again, but this time only a few flies were killed. What explanation can you give for these events?

These orange trees have been attacked by locusts

Part of a large swarm of locusts

Colorado beetles on a potato plant

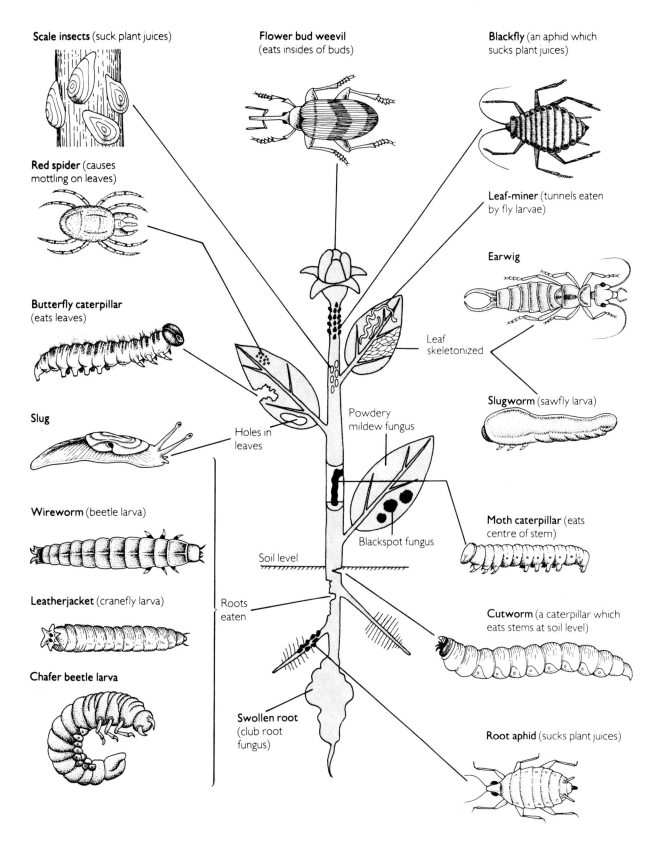

Scale insects (suck plant juices)

Red spider (causes mottling on leaves)

Butterfly caterpillar (eats leaves)

Slug

Wireworm (beetle larva)

Leatherjacket (cranefly larva)

Chafer beetle larva

Flower bud weevil (eats insides of buds)

Holes in leaves

Soil level

Roots eaten

Swollen root (club root fungus)

Blackfly (an aphid which sucks plant juices)

Leaf-miner (tunnels eaten by fly larvae)

Earwig

Leaf skeletonized

Slugworm (sawfly larva)

Powdery mildew fungus

Blackspot fungus

Moth caterpillar (eats centre of stem)

Cutworm (a caterpillar which eats stems at soil level)

Root aphid (sucks plant juices)

Fig. 1 Some pests which attack vegetables

Topic 3 exercises

An under-nourished child (see exercise 1)

1 Study the photograph of the child who has never had enough to eat.

a) Make a list of the visible features which you think are caused by starvation.

b) How do you think the child's energy, growth, resistance to disease, ability to heal wounds, and resistance to cold weather will have been affected by his poor diet?

c) If you were able to help nourish the child, what foods would you give him to provide energy, build up his body, and increase his resistance to disease?

2 Cut out pictures of food from women's and teenagers' magazines. Use these pictures to make a large wall chart showing foods which are a good source of carbohydrate, protein, fat and oil, and vitamins.

3 Study each of the following meals and describe what there is too much of, and what there is too little of (i.e. in what way each is an unbalanced meal). How could each meal be improved?

a) Steamed rice and boiled vegetables.

b) Fried fish and chips.

c) Sandwiches made with white bread, butter, tomatoes, and lettuce.

d) Poached egg and grilled hamburger.

4 In 1912, before anyone knew about vitamins, a scientist performed two experiments using two groups of young rats:

1st Experiment

Group A rats were fed highly purified protein, carbohydrate, fat, minerals, and water. The food contained no vitamins although the scientist did not know this.

Group B rats were fed the same diet but with a *few drops* of milk added.

Group A rats stopped growing and lost weight. Group B rats grew steadily and gained weight.

2nd Experiment

The milk was removed from Group B's diet and added to Group A's diet. Group A rats now gained weight and Group B rats stopped growing and lost weight.

These results told the scientist *two* important things about milk. What are these two things? Why was the second experiment necessary?

5 If iodine solution is poured onto starch, the starch changes to a blue-black colour. Try putting a drop of iodine solution onto bread, meat, the cut surface of a potato, a bean, a maize grain, and boiled egg white. What conclusions do you reach about these foods?

6 a) What do digestive enzymes do to food?

b) Three test tubes were prepared to investigate the effect of stomach digestive enzyme on boiled egg white.

Tube 1	Tube 2	Tube 3
Mashed egg white	Mashed egg white	Mashed egg white
Enzyme	Enzyme	Enzyme
Kept at 37 °C	Kept at 0 °C	Kept at 100 °C

After 15 minutes the egg in tube 1 was completely dissolved, but in the other tubes it was unchanged after 1 hour. What does this tell you about stomach enzymes?

c) Two more test tubes were prepared as follows.

Tube 1	Tube 2
Mashed egg white	Large lump of egg white
Enzyme	Enzyme
Kept at 37 °C	Kept at 37 °C

The egg in tube 1 dissolved in 15 minutes, but in tube 2 the egg took 45 minutes to dissolve. What does this tell you about the importance of chewing food thoroughly?

Energy from food

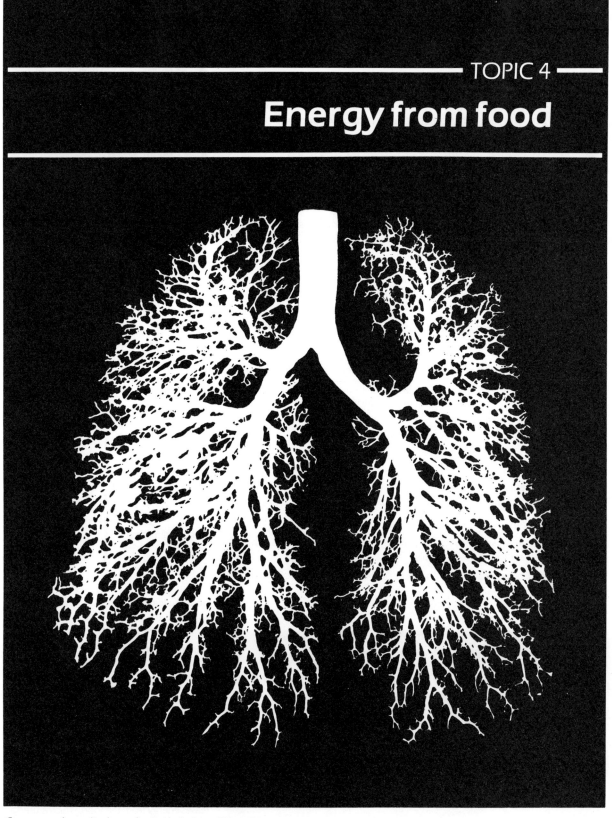

Gases are exchanged in these tubes inside the lungs. This exchange is important in the process of obtaining energy from food.

4.1 Respiration

In most organisms the energy for life comes from a chemical reaction between food and oxygen. This reaction is called cellular respiration because it occurs in every cell of the body.

Cellular respiration

During cellular respiration food is combined with oxygen and broken down into carbon dioxide gas and water. This releases energy. The body uses this energy in the ways described in Figure 1. Carbon dioxide gas and water are waste products of respiration, and are released from the body.

This photo shows part of a dogfish gill (magnified)

Respiratory organs

Animals use respiratory organs to breathe in oxygen for respiration, and to breathe out the carbon dioxide produced by respiration. In other words, respiratory organs _exchange_ carbon dioxide gas for oxygen gas, a process called **gaseous exchange**.

The gills of fish and prawns are examples of respiratory organs which exchange carbon dioxide for oxygen in water (Figs. 2 and 3). The tracheal system of insects (Fig. 4) and the human lungs are respiratory organs which exchange carbon dioxide for oxygen in air.

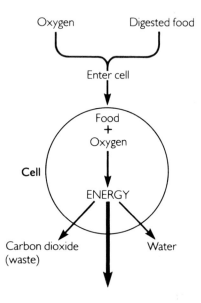

Exercises

1 What two substances react together during respiration, and what waste product is released?

2 What use do living things make of the energy released during respiration?

3 What is gaseous exchange?

4 Study Figures 2 and 3. What is the difference between the way a fish and a prawn make water flow over their gills?

5 What is a spiracle? How does an insect pump air in and out of its tracheal system?

Body-building: Manufacture of proteins and other materials needed to make new cells for growth and repair

Muscular work: Contraction of muscles which move the body, heart muscle, gut muscle, etc.

Chemical work: In the liver, kidneys, brain, and nerves (movement of nerve impulses) and movement of chemicals in and out of cells.

A dolphin comes to the surface to breathe through its nostril

Fig. 1 Diagram of respiration Food is combined with oxygen and broken down into carbon dioxide gas and water. This releases energy from the food, which is used as shown. Carbon dioxide is released as a waste product.

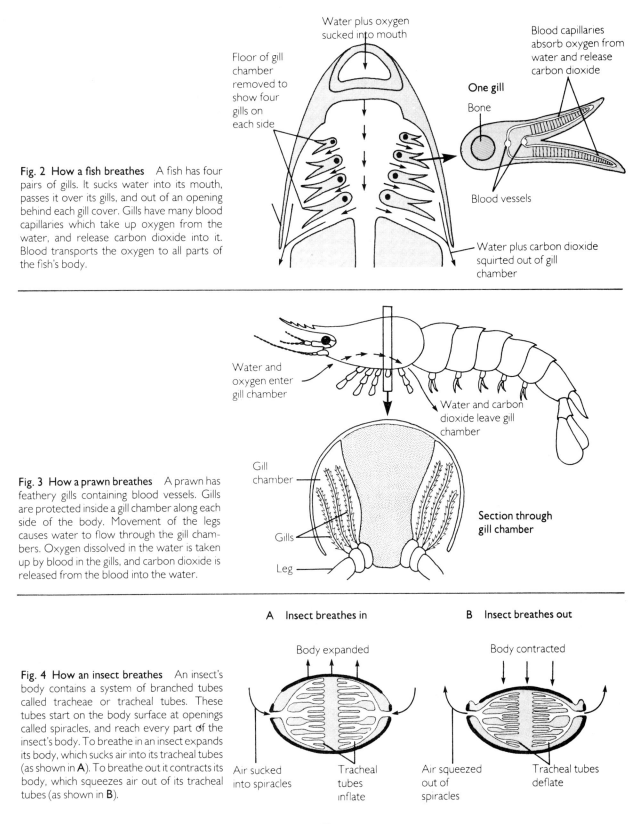

Fig. 2 How a fish breathes A fish has four pairs of gills. It sucks water into its mouth, passes it over its gills, and out of an opening behind each gill cover. Gills have many blood capillaries which take up oxygen from the water, and release carbon dioxide into it. Blood transports the oxygen to all parts of the fish's body.

Water plus oxygen sucked into mouth

Floor of gill chamber removed to show four gills on each side

Blood capillaries absorb oxygen from water and release carbon dioxide

One gill

Bone

Blood vessels

Water plus carbon dioxide squirted out of gill chamber

Fig. 3 How a prawn breathes A prawn has feathery gills containing blood vessels. Gills are protected inside a gill chamber along each side of the body. Movement of the legs causes water to flow through the gill chambers. Oxygen dissolved in the water is taken up by blood in the gills, and carbon dioxide is released from the blood into the water.

Water and oxygen enter gill chamber

Water and carbon dioxide leave gill chamber

Gill chamber

Gills

Leg

Section through gill chamber

Fig. 4 How an insect breathes An insect's body contains a system of branched tubes called tracheae or tracheal tubes. These tubes start on the body surface at openings called spiracles, and reach every part of the insect's body. To breathe in an insect expands its body, which sucks air into its tracheal tubes (as shown in **A**). To breathe out it contracts its body, which squeezes air out of its tracheal tubes (as shown in **B**).

A   Insect breathes in

Body expanded

Air sucked into spiracles

Tracheal tubes inflate

B   Insect breathes out

Body contracted

Air squeezed out of spiracles

Tracheal tubes deflate

4.2 Human respiratory organs

Human lungs have an internal surface area equivalent to a singles tennis court and contain enough blood vessels to reach from London to New York.

Structure of the respiratory organs

Figure 2 illustrates the structure of the human respiratory organs.

Voice box (larynx) This is a box with walls made of gristle (Fig. 2A). It contains thin sheets of skin called the *vocal cords*. When you speak air passes over these cords in a way which makes them vibrate and produce sound.

Wind-pipe (trachea) The wind-pipe is a tube which connects the mouth with the lungs. Its walls are stiffened with rings of cartilage, which hold it open so that you can breathe freely at all times (Fig. 2A).

Lungs At its lower end the wind-pipe divides into two tubes, one for each lung. Inside the lungs these tubes divide like the branches of a tree, forming smaller and smaller tubes. The smallest and narrowest of these *bronchial tubes* end in a cluster of tiny bubbles, or *air sacs*, 0.2 mm in diameter (Fig. 2B).

There are 300 million of these air sacs in a set of human lungs, and each is covered with a net of blood capillaries (Fig. 2C). This arrangement makes lungs similar in appearance to large, pink sponges.

Exchange of gases in the lungs

The air sacs of the lungs fill with air when you breathe in and empty when you breathe out. Oxygen from the air dissolves in a film of water lining each air sac and passes through the walls of the air sac into a blood vessel on the other side (Fig. 1). Red cells in the blood absorb the oxygen and carry it to all parts of the body. Blood absorbs carbon dioxide from the body and carries it back to the lungs. Here, carbon dioxide passes from the blood into the air sacs and is breathed out of the body.

Exercises

1 What are the trachea, larynx, vocal cords, and bronchial tubes for?

2 How is the voice produced?

3 What is the function of the rings of gristle in the wind-pipe walls?

4 How does oxygen get from the atmosphere to cells in your toes? How does carbon dioxide get from your toes back to the atmosphere?

5 In which parts of the lungs does gaseous exchange occur?

6 Which part of your blood carries oxygen around your body?

7 How does blood entering the lungs from the body differ from blood leaving the lungs?

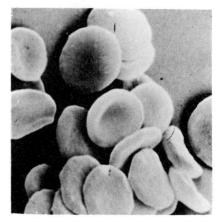

Red blood cells carry oxygen around the body

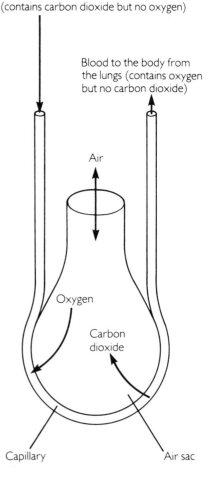

Blood from the body to the lungs (contains carbon dioxide but no oxygen)

Blood to the body from the lungs (contains oxygen but no carbon dioxide)

Air

Oxygen

Carbon dioxide

Capillary

Air sac

Fig. 1 Exchange of gases in the air sacs of the lungs

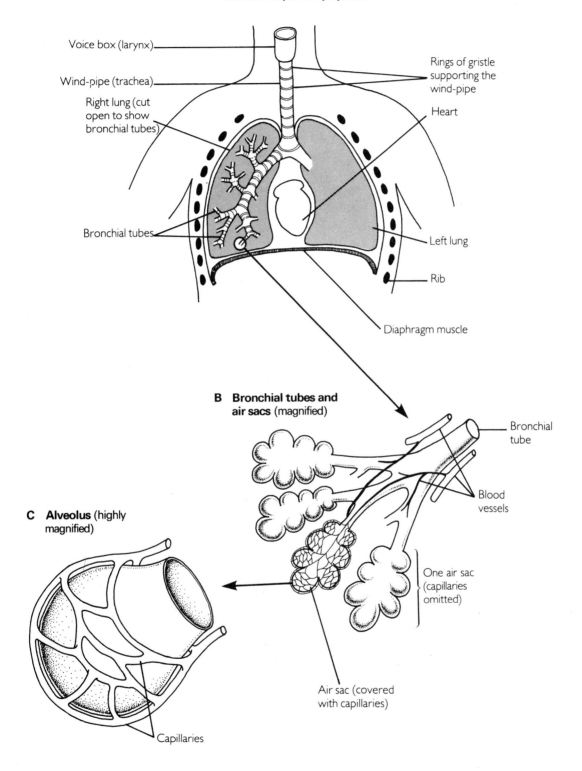

A Human respiratory system

Voice box (larynx)

Wind-pipe (trachea)

Right lung (cut open to show bronchial tubes)

Bronchial tubes

Rings of gristle supporting the wind-pipe

Heart

Left lung

Rib

Diaphragm muscle

B Bronchial tubes and air sacs (magnified)

Bronchial tube

Blood vessels

One air sac (capillaries omitted)

C Alveolus (highly magnified)

Air sac (covered with capillaries)

Capillaries

Fig. 2 Parts of the human respiratory system

4.3 Breathing

How many times do you breathe in and out in one minute when you are relaxed, and after exercise? Can you account for the difference?

Relaxed breathing

When you are sitting in a relaxed position, air is pumped in and out of your lungs by your **diaphragm muscle**. The diaphragm is a sheet of muscle which forms the floor of the space in your chest which contains your lungs (Figs. 1 and 2).

Breathing in   Just before breathing in the diaphragm is curved upwards and its muscle fibres are relaxed (Fig. 1A). To breathe in, the diaphragm contracts and becomes flatter in shape (Fig. 1B). This sucks air down the wind-pipe and into the lungs, inflating them.

Breathing out   The diaphragm now relaxes and returns to its original curved shape. This pushes air out of the lungs (Fig. 1A).

Deep breathing

During exercise muscles require far more oxygen and produce far more carbon dioxide than when the body is relaxed. Breathing automatically becomes faster and deeper during exercise and for a short time after-wards, in order to supply the extra oxygen and remove the extra carbon dioxide.

During deep breathing you use your diaphragm muscle *and* muscles between the ribs. The diaphragm muscle acts as described above. The muscles between the ribs contract and lift the whole rib cage (Figs. 1A and 2A). Take a deep breath and feel this happen. This upward movement of the rib cage sucks far more air into the lungs than contraction of the diaphragm alone.

To breathe out the diaphragm and rib-lifting muscles relax. But in addition, another set of muscles between the ribs contract and pull the rib cage downwards (Figs. 1B and 2B). Blow air forcibly from your lungs and feel these muscles in action.

Exercises

1   Where is the diaphragm muscle situated in the body?
2   a) What is the shape of the diaphragm when it is relaxed and when it is contracted?
    b) What happens to the lungs when the diaphragm contracts, and when it relaxes?
3   a) Which gas does the body use more of during exercise and which gas does it produce more of?
    b) Describe how extra air is sucked into the lungs during deep breathing.
    c) Which muscles do you use to blow up a balloon?

A   Impure air breathed out

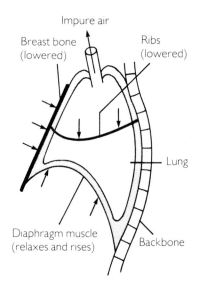

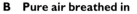

B   Pure air breathed in

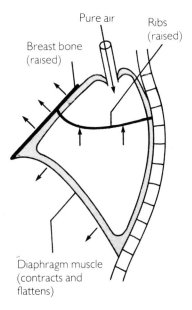

Fig. 1 Side view of the chest during deep breathing

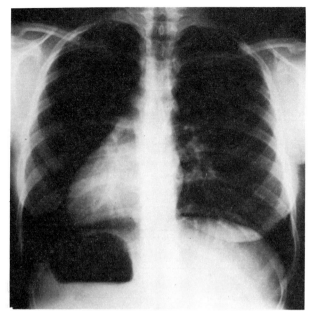

X-ray of a healthy pair of lungs. Compare them with those on page 147

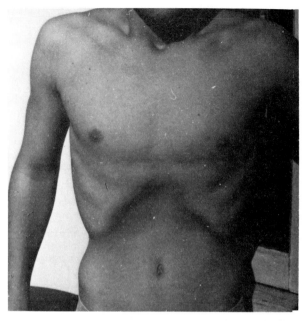

During deep breathing the rib cage moves up and air is sucked into the lungs

A   Breathing in   B   Breathing out

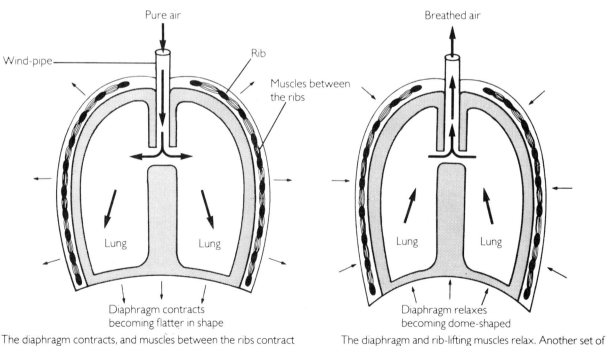

The diaphragm contracts, and muscles between the ribs contract lifting the rib cage

The diaphragm and rib-lifting muscles relax. Another set of muscles between the ribs contract pulling the rib cage downwards

Fig. 2 Front view of the chest during deep breathing

Topic 4 exercises

1   Lime water is a chemical which changes to a milky-white colour when carbon dioxide gas is bubbled through it. Study Figure 1. A student breathed air into tube A of this apparatus so that his breath bubbled through the lime water. After 10 seconds the lime water became milky in colour. The test tube was cleaned and refilled with lime water. The student then sucked at tube B, which sucked classroom air through the lime water. It became milky after 3 minutes. The tube was cleaned and refilled again. The student opened a window and sucked fresh, clean outdoor air through the lime water. It had not become milky after 5 minutes so the exercise was stopped. What conclusions can you draw from these results?

3   Study the chart below. What does it tell you about how air is changed inside the lungs, and about the effects of exercise on these changes?

	Unbreathed air	Breathed air from a sleeping man	Breathed air from a man cycling
Nitrogen	78%	78%	78%
Oxygen	21%	17%	12%
Carbon dioxide	0.03%	4%	9%
Water vapour	Variable	Saturated	Saturated

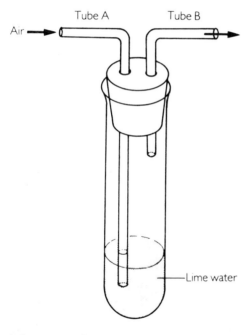

Fig. 1 See exercise 1

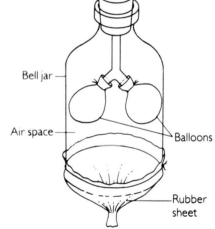

Fig. 2 See exercise 4

2   Obtain a large plastic bag (like a wastebin liner). Take a deep breath and inflate the bag (try to breathe all the air out of your lungs). Tie a knot in the bag to prevent air escaping. Estimate the volume of the bag (and so of your lungs) by pushing the bag into a round bucket. Press the bag down with a piece of stiff card cut to size and mark the side of the bucket at a point level with the card. Remove the card and bag and fill the bucket with water up to the mark. How much water is needed? Is this a measure of the *total* volume of your lungs?

4   Study Figure 2
  a) What part of the human body do the glass tube, bell jar, balloons, and rubber sheet represent?
  b) What will happen to the balloons when the rubber sheet is pulled down, and when it is pushed up?
  c) How is this similar to, and different from, the way your lungs are inflated and deflated?
  d) What method of lung inflation is not illustrated by this apparatus?

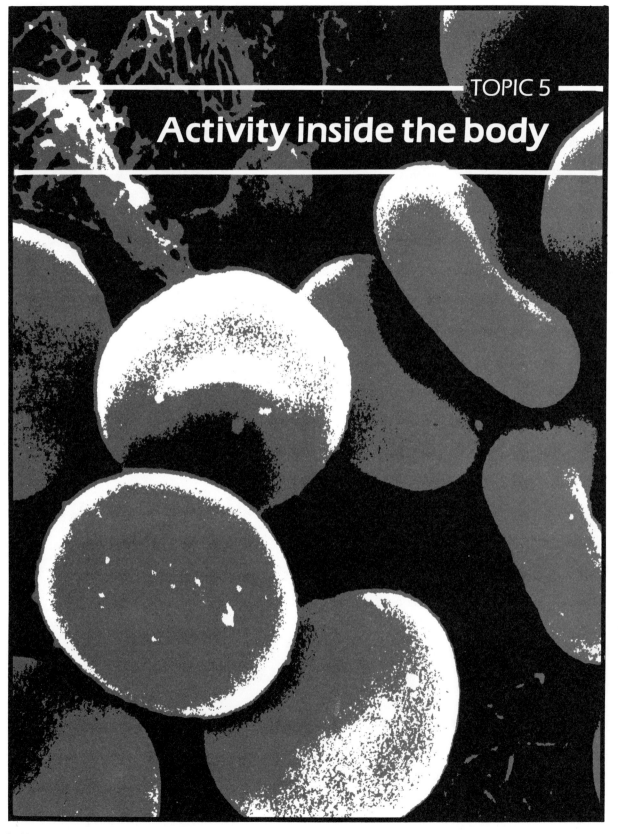

Activity inside the body

Red blood cells carry oxygen around the body

5.1 Transport systems

Animals are made up of millions of cells, each of which must have a continuous supply of food and oxygen, and must be able to get rid of wastes. These needs are satisfied by a transport system.

Transport in very small organisms

Microscopic organisms like *Amoeba* do not need an elaborate transport system. Their bodies are so small that a process called **diffusion** is sufficient to transport substances into, around, and out of their bodies. Diffusion is the movement of substances from where they are plentiful to where they are scarce (Fig. 1).

Transport in plants

Ferns and seed plants have a transport system made up of tubes which extend from the roots, through the stem, and into the leaves where they can be seen as a pattern of veins (Fig. 2).

Transport in animals

In animals a muscular pump, the heart, sends a liquid called blood on a circular route around the body. This arrangement is known as a *circulatory system*.

Open circulatory system   This is found in insects and other arthropods. Here, blood travels only part of the way around the body in blood vessels. It completes the circuit by travelling through the open spaces inside the body. In these open spaces it comes into direct contact with muscles, nerves, glands, and the organs of the body, before returning to the heart (Fig. 3).

Closed circulatory system   Vertebrates (fish, amphibia, reptiles, birds, and mammals) have a closed circulatory system, so-called because blood travels all the way around the body enclosed in blood vessels. This type of transport system is described in the next three Units.

Exercises

1 What is diffusion?

2 What substances diffuse into and out of an *Amoeba*?

3 What substances travel from the roots to the leaves of plants, and what substance travels from the leaves to other parts of a plant?

4 What part does the heart play in the circulation of blood?

5 What is the difference between an open and a closed circulatory system?

6 Sort the following into those with an open circulatory system and those with a closed circulatory system:
cockroach, human, frog, crab, penguin, woodlouse.

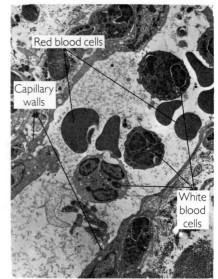

Red and white blood cells inside a blood vessel ($\times$ 4500)

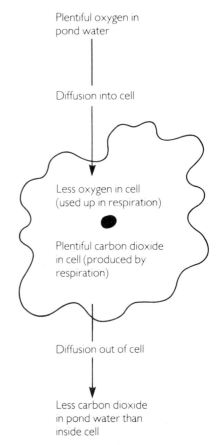

Fig. 1 Transport by diffusion in Amoeba

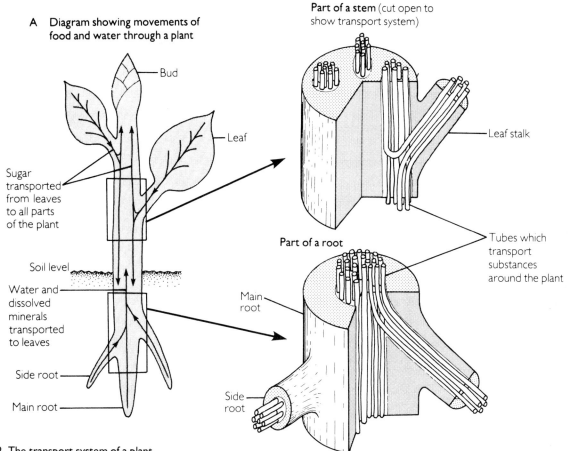

A   Diagram showing movements of
    food and water through a plant

Bud

Leaf

Sugar
transported
from leaves
to all parts
of the plant

Soil level

Water and
dissolved
minerals
transported
to leaves

Side root

Main root

Part of a stem (cut open to
show transport system)

Leaf stalk

Part of a root

Main
root

Side
root

Tubes which
transport
substances
around the plant

Fig. 2 The transport system of a plant

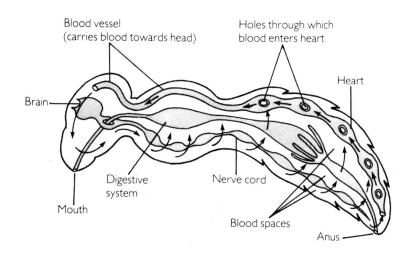

Blood vessel
(carries blood towards head)

Holes through which
blood enters heart

Heart

Brain

Digestive
system

Nerve cord

Blood spaces

Mouth

Anus

**Fig. 3 The open circulation of an
insect** Blood completes its circuit of the
body by flowing through the open spaces
between tissues and organs.

→ → Direction of blood flow

75

5.2 Human blood

Blood is alive. One drop contains about 5 million cells. Blood carries many substances around your body, soaks up wastes, contains cells which eat germs, and does 'puncture repairs' by clotting in wounds.

What is blood?

Blood consists of four main things: *red blood cells, white blood cells,* and *platelets,* all floating in a watery liquid called *plasma.*

Red blood cells

Red blood cells carry oxygen from the lungs to all parts of the body. They contain a chemical called *haemoglobin* which makes them red and which enables them to carry oxygen.

Adult humans contain about five and a half litres of blood—in this, there are about thirty million million red blood cells.

The shape of a red blood cell is a bi-concave disc (Fig. 1 and p. 68). Red blood cells are very small—five thousand of them stacked on top of one another would only make a pile one centimetre high.

After four months of use, red cells are worn out and destroyed by the spleen. But some of the chemicals in them are re-used to make new cells, which are produced at a rate of two million every second!

White blood cells

There are about 5000 times more red cells than white cells but white cells are about twice as big as red cells. This is shown in Figure 1.

White cells protect you by destroying germs that enter your body, in two different ways:
1 Some white cells 'eat' germs by engulfing and then destroying them (Fig. 2).
2 Some white cells make chemicals called *antibodies.* These kill germs, and change the poisonous chemicals produced by germs into harmless substances. Some antibodies remain in the blood for months or even years after they have helped you recover from a disease. While they are there, they help to stop your catching the disease again.

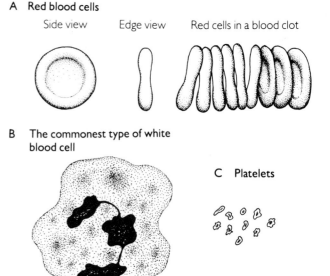

A Red blood cells

Side view   Edge view   Red cells in a blood clot

B The commonest type of white blood cell

C Platelets

Nucleus (with three lobes)

Fig. 1 Blood cells Red cells are bi-concave discs, and are so small that it would take 5000 stacked like coins to make a pile 1 cm high. White cells are twice the size of red ones, and can change shape like an *amoeba*. Platelets are tiny pieces of cells.

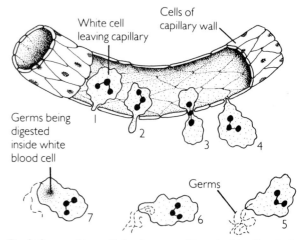

White cell leaving capillary

Cells of capillary wall

Germs being digested inside white blood cell

Germs

Fig. 2 Some white cells 'eat' germs They do this either in the bloodstream or after squeezing between cells of a capillary wall (**1, 2, 3,** and **4**) to attack germs among cells. They engulf (swallow) germs and digest them (**5, 6,** and **7**).

Platelets

Platelets are tiny pieces of cells released into the blood from the bone marrow (Fig. 1). Platelets are part of the body's puncture repair kit—they help prevent loss of blood from wounds. When skin is cut it bleeds for a while, washing dirt and germs from the wound. Soon, platelets form a pad of fibres over the wound. Red cells are trapped amongst these fibres. As the red cells dry out, they make a solid plug which stops the bleeding, and keeps out dirt and germs until the wound heals (Fig. 3).

Plasma

In blood, red and white cells and platelets float in a liquid called plasma. Plasma is water in which food, carbon dioxide, and other important substances are dissolved. It carries these many substances around the body (Fig. 4).

It carries the red cells which take oxygen to all parts of the body. It carries the white cells which destroy germs. It carries the chemicals which help platelets to act as a 'puncture repair kit'. It carries food from the gut to the liver, and from the liver to other parts of the body. It carries carbon dioxide from the body back to the lungs. It carries a waste product called urea from the liver to the kidneys.

A **blood clot** showing red cells trapped in a mesh of fibres (✕ 2400)

Exercises

1 What is the liquid part of blood called? What is dissolved in this liquid?

2 How do red and white cells differ in shape and size?

3 What is haemoglobin, where is it found, and what are its functions?

4 Describe the functions of white cells.

5 How do platelets help stop you from bleeding to death when you cut yourself?

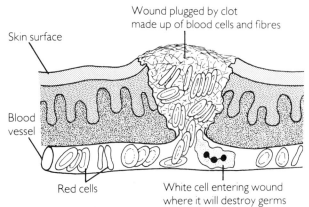

Fig. 3 The body's puncture repair kit Platelets form a pad of fibres over a wound. Red cells trapped in the fibres plug the wound until it heals (see photograph).

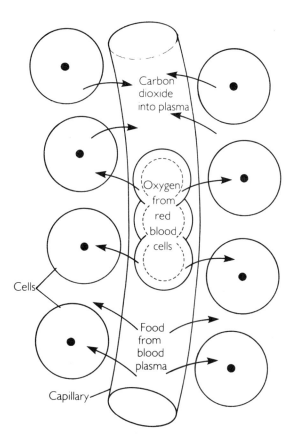

Fig. 4 The main functions of plasma and red cells Plasma transports food from the gut to the liver, and from the liver to the body. Plasma also carries carbon dioxide from the body to the lungs, and urea (a waste produced in the liver) to the kidneys for excretion. Plasma contains chemicals concerned with blood clotting. Red cells transport oxygen from the lungs to all body cells.

5.3   The human heart

The heart weighs 300 g. In an average life-time it beats 2 500 million times and pumps 340 million litres of blood.

Heart beats

The heart is a pump with walls made of muscle. When this muscle contracts it squeezes blood out of the heart and around the body. When its muscle relaxes the heart fills with blood returning from its journey around the body. One complete contraction (Figs. 1B and C) and one relaxation of the heart make up a *heart beat*. Your heart beats about 75 times a minute when you are resting, rising to about 140 times a minute during exercise.

The space inside the heart is divided into four compartments (Figs. 1, 2, and 3). The two top compartments are called *atria* (singular: atrium). The atria have thin muscular walls. The two lower compartments are called *ventricles* and have much thicker muscular walls.

Blood returning from a journey around the body enters the atria as they relax. The atria then contract, forcing blood down into the ventricles. Next the ventricles contract, pumping blood out of the heart (Fig. 1).

Heart valves

Blood can only flow through the heart in *one* direction. This is because blood flow is controlled by valves.

Valves between atria and ventricles   These valves consist of flaps of skin held in place by tendons of tough fibre (Fig. 3). When the atria contract blood pushes these flaps open, like doors, and flows past them into the ventricles (Fig. 1B). When the ventricles contract, blood pushes the flaps together again (Fig. 1C). Therefore, blood can only flow out of the heart.

Pocket valves   These are situated at the exits from the heart. They consist of three pockets in the walls of the main blood vessels (Fig. 3). Pocket valves open as blood flows past them from the ventricles (Fig. 1C). But when the ventricles relax the pockets fill with blood and close the exits from the heart, so that blood cannot reverse its direction and rush back into the ventricles (Fig. 1B).

Exercises

1   Name the four compartments of the heart.

2   What is a heart beat?

3   What is the difference between the walls of the atria and of the ventricles? Can you account for this difference?

4   Describe the valves situated between the atria and ventricles. When do these valves open and when do they close?

5   Describe the valves at the exit from the heart. When do they open and when do they close?

A   Between heart beats the heart muscle is relaxed and the heart fills with blood

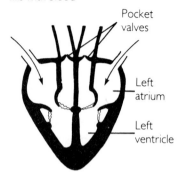

B   Atria contract and pump blood down into the ventricles

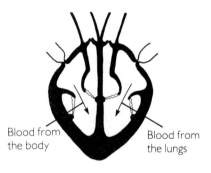

C   Ventricles contract pumping blood out of the heart. Atria fill again

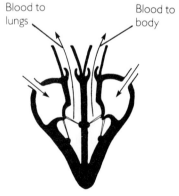

Fig. 1 Diagram of how the heart pumps blood

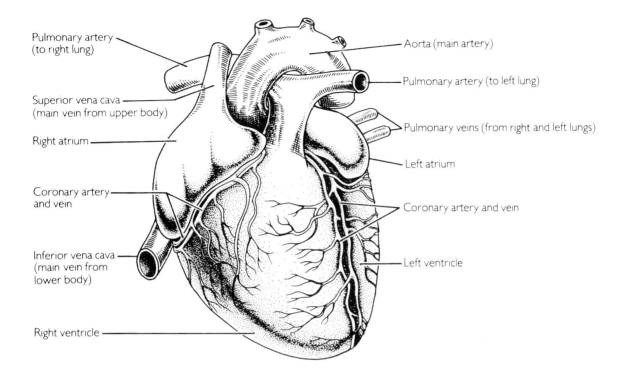

Pulmonary artery
(to right lung)

Aorta (main artery)

Pulmonary artery (to left lung)

Superior vena cava
(main vein from upper body)

Pulmonary veins (from right and left lungs)

Right atrium

Left atrium

Coronary artery
and vein

Coronary artery and vein

Inferior vena cava
(main vein from
lower body)

Left ventricle

Right ventricle

Fig. 2 The human heart

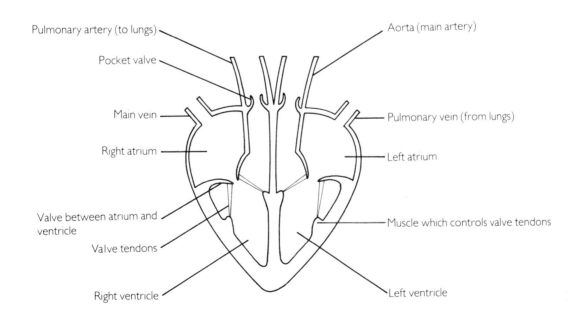

Pulmonary artery (to lungs)

Aorta (main artery)

Pocket valve

Main vein

Pulmonary vein (from lungs)

Right atrium

Left atrium

Valve between atrium and
ventricle

Muscle which controls valve tendons

Valve tendons

Right ventricle

Left ventricle

Fig. 3 Diagram showing the parts of the heart

5.4 Blood vessels and circulation of the blood

There are so many blood vessels in the human body that if everything else were dissolved away the body outline and all its organs would still be visible owing to the mass of vessels left behind.

Blood vessels

The heart pumps blood into vessels called **arteries**. Arteries have thick muscular walls able to withstand the high pressure of blood passing through them.

Arteries divide up into narrower and narrower vessels until they form **capillaries**. Some capillaries are narrower than red blood cells so they are bent out of shape as the blood cells pass through.

Capillary walls are so thin that they leak: liquid passes through them from the blood, carrying dissolved food and oxygen to all the cells of the body. This liquid eventually drains back into the bloodstream bringing with it carbon dioxide and other waste from the cells.

Blood travels back to the heart through vessels called **veins**. Veins have thinner walls than arteries, carry blood at lower pressure, and contain pocket valves which ensure that blood flows in only one direction.

Circulation of blood

Humans and all other mammals have a circulatory system made up of two parts joined at the heart. One of these parts serves the lungs and the other serves the rest of the body.

Look at Figures 1 and 2 as you read this description of blood circulation. Circulation to the lungs begins at the right ventricle. Blood flows from here through the **pulmonary artery** to the lungs where it loses carbon dioxide and picks up oxygen. Blood then flows through the **pulmonary vein** to the left atrium of the heart.

Next, blood is pumped from the left atrium to the left ventricle which then pumps it at high pressure through the **main artery** to all parts of the body *except* the lungs. Here blood gives up its oxygen and picks up carbon dioxide, before returning through the **main vein** to the right atrium.

The pulse

Each time the heart pumps blood into the main artery it causes a ripple to move along the walls of the artery away from the heart. This ripple is called the *pulse*. It can be felt with the fingertips in arteries of the wrist (see p. 86) and on either side of the wind-pipe.

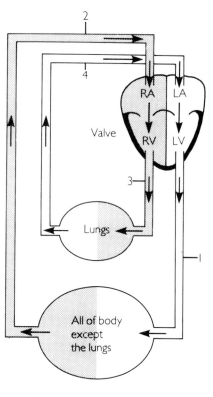

Key:

1	Main artery
2	Main vein
3	Pulmonary artery
4	Pulmonary vein
RA	Right atrium
RV	Right ventricle
LA	Left atrium
LV	Left ventricle

▨▨▨ Blood without oxygen but with carbon dioxide

Fig. 1 Diagram showing the two parts of the human (and mammal) circulatory system One part serves the lungs, the other serves the body except the lungs (see exercise 3).

Exercises

1   Sort the following statements into those which describe arteries, veins, or capillaries:

    a) Have thin walls and pocket valves.

    b) Carry blood away from the heart at high pressure.

    c) Extremely narrow.

    d) Carry blood back to the heart.

    e) Their walls are so thin that fluid leaks out through them.

    f) A ripple called the pulse passes along them.

2   Which ventricle pumps blood to the lungs, and which pumps blood to the remainder of the body?

3   Which blood vessel in Figure 1 is described by each of the following sentences?

    a) Contains blood full of oxygen at low pressure.

    b) Contains blood full of carbon dioxide at high pressure.

    c) Contains blood full of carbon dioxide at low pressure.

    d) Contains blood full of oxygen at very high pressure.

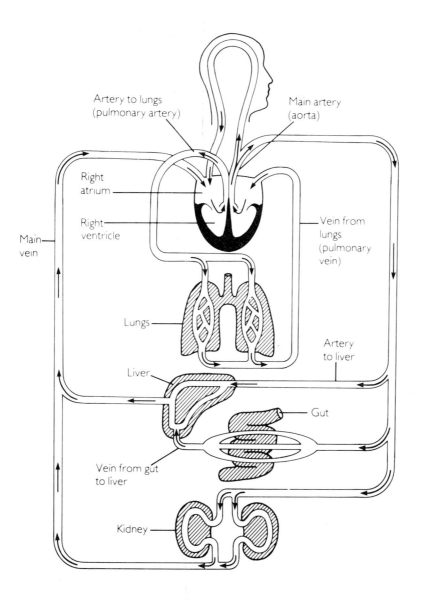

Fig. 2 Diagram of the main blood vessels

5.5 Heart disease

Heart disease kills more people than any other illness. But it is decreasing because of modern discoveries about its causes, treatment, and prevention.

Clogged arteries

The main cause of heart disease is the build-up of a fatty substance called *cholesterol* inside arteries (Fig. 1). This makes arteries narrower and eventually clogs them up altogether. The coronary arteries supply heart muscle with food and oxygen (see Fig. 2, p. 79). If cholesterol clogs the coronary arteries, the heart becomes weaker and may stop beating. This is called heart failure.

Treatment of heart failure

Electric shock treatment   The heart has a built-in electrical system, called the *pacemaker*, which keeps the heart muscles working by sending tiny, regular electric currents through them. During one type of heart failure the pacemaker stops working so that, instead of beating regularly, the heart flutters uselessly. A regular heart beat can be restored by shocking the heart back to life. Electrical contacts are placed on either side of the chest and an electric shock of up to 200 volts is passed across them.

Artificial pacemakers   If the heart's pacemaker becomes unreliable an artificial one can be fitted. An artificial pacemaker is a battery-operated device which sends regular electric shocks through wires connected directly to the heart. These shocks cause the heart muscles to contract. Modern artificial pacemakers have batteries which last five years, and the device is small enough to be inserted surgically under the skin somewhere near the heart (see photograph).

Transplant surgery

The first time a diseased human heart was replaced with a healthy one from another person (donor) the patient lived for only 18 days. Today, survival rates are 66 per cent after one year and 50 per cent after five years.

The transplant operation   The patient is first connected to a heart–lung machine. This collects blood before it enters the heart, charges it with oxygen, and pumps it back around the patient's body. The donor 'new' heart is packed in crushed ice until needed. Most of the patient's heart is removed leaving only the back wall of the upper chambers (atria). The donor heart is then sewn into place. Figure 2 shows an alternative operation called the piggyback technique.

Rejection of the transplant

The body's *immune system* can detect and destroy any foreign material, such as germs which enter it. Unfortunately, the immune system often treats transplants as if they were germs and destroys (rejects) them. Rejection is less likely if organs come from a close relative.

There are now many drugs which reduce the risk of transplant rejection by damping down the immune system. But, as a result, the patient is left wide open to infection, especially from pneumonia.

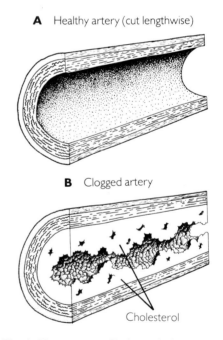

A   Healthy artery (cut lengthwise)

B   Clogged artery

Cholesterol

Fig. 1 The hidden enemy   Cholesterol clogs arteries, slows blood flow, and may stop it altogether. Clogged arteries are more likely to occur in people who eat food rich in animal fats (e.g. dairy produce), who take little exercise, who smoke, and whose lives are full of stress (tension, anxiety, fear, etc.).

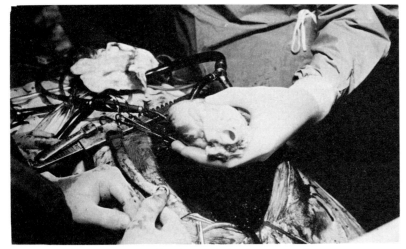

A heart transplant operation The chest cavity has been opened and the donor heart is being held above it.

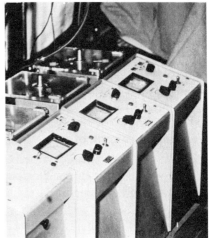

A heart-lung machine

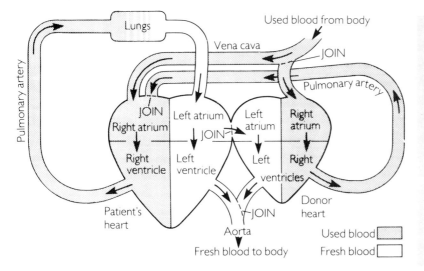

Fig. 2 Piggyback heart transplant The patient's heart is left intact. The donor heart is connected to it as shown. The donor heart helps rather than replaces the patient's heart.

An artificial pacemaker is usually placed under the skin on one side of the chest. It sends electric shocks to the heart to keep the heart muscles working.

Exercise

Between 1963 and 1975 coronary heart disease among Americans between 37 and 75 years of age fell by a quarter. During this time tobacco use fell by 22 per cent, consumption of vegetable fats increased by 44 per cent, and less of each of the following foods was consumed:

Milk and cream	down	19%
Butter	down	31%
Eggs	down	12%
Animals fats	down	56%

Use these facts to write a summary of how you can help yourself to avoid coronary heart disease.

5.6 Temperature control and excretion

If it were not for non-stop activity in the liver, lungs, kidneys, and skin, the body would quickly be poisoned by its own wastes, and its temperature would shoot up and down out of control.

Temperature control

Humans, other mammals, and birds can automatically keep their bodies at a constant temperature. When you are healthy your temperature is 37 °C. When you are too hot blood vessels and 2 million sweat glands in the skin quickly get rid of the excess heat as follows.

Skin blood vessels These become wider and carry extra, over-heated blood to the body surface where it passes into the air (Fig. 1).

Sweat glands These release water (sweat) onto the skin's surface where it evaporates and cools the body. In hot weather you can lose up to 12 litres of water and 30 g of salt in a day through sweating. In these conditions your kidneys absorb less water from the blood, but it is important to drink more, and put extra salt on your food to make up for the loss.

In cold weather, you put on extra clothing, turn on a heater, etc., and, in addition, your body automatically generates extra heat. The liver in particular makes more heat by respiring faster, and muscles make more heat by shivering. The skin blood vessels contract to reduce heat loss by radiation, and you do not sweat (Fig. 2).

Excretion

Excretion is the removal from the body of waste and unwanted substances. The lungs excrete water vapour and carbon dioxide, the wastes produced by respiration. The liver excretes bile, which contains unwanted chemicals released when old red blood cells are destroyed by the spleen. One of the most important examples of excretion is the removal of urea from the body by the kidneys. Urea is a chemical formed in the liver when it breaks down the unwanted parts of protein foods.

Kidneys and excretion

Blood is filtered in the kidneys to remove urea, and water, sugar, and salts which are unwanted. These filtered substances form a liquid called *urine*, which drains out of the kidneys into the bladder (Fig. 3). The bladder is emptied when you use the lavatory.

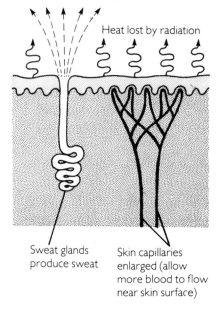

Heat lost by evaporation of sweat

Heat lost by radiation

Sweat glands produce sweat

Skin capillaries enlarged (allow more blood to flow near skin surface)

Fig. 1 In hot weather the skin loses heat by evaporation of sweat and radiation of heat.

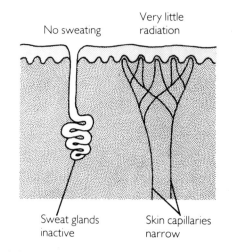

No sweating

Very little radiation

Sweat glands inactive

Skin capillaries narrow

Fig. 2 In cold weather the skin retains heat by reducing sweating and radiation.

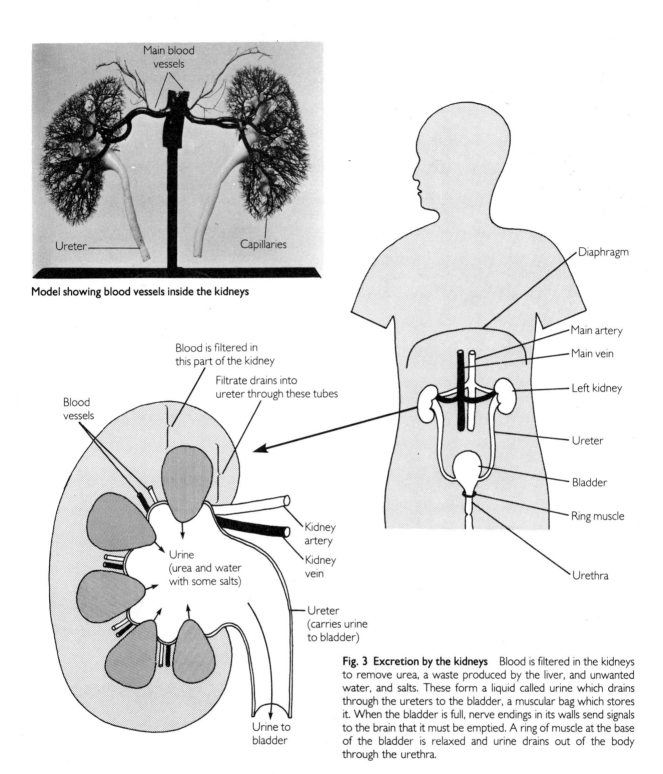

Main blood vessels

Ureter

Capillaries

Model showing blood vessels inside the kidneys

Diaphragm

Main artery

Main vein

Left kidney

Ureter

Bladder

Ring muscle

Urethra

Blood is filtered in this part of the kidney

Filtrate drains into ureter through these tubes

Blood vessels

Kidney artery

Kidney vein

Urine (urea and water with some salts)

Ureter (carries urine to bladder)

Urine to bladder

Fig. 3 Excretion by the kidneys Blood is filtered in the kidneys to remove urea, a waste produced by the liver, and unwanted water, and salts. These form a liquid called urine which drains through the ureters to the bladder, a muscular bag which stores it. When the bladder is full, nerve endings in its walls send signals to the brain that it must be emptied. A ring of muscle at the base of the bladder is relaxed and urine drains out of the body through the urethra.

Exercises

1 Which animals have a constant body temperature? Can you think of any advantages of having this ability?
2 What happens in your skin when you are too hot?
3 How is urine produced and what does it consist of?

4 What is excretion? What are the main excretory organs, and what does each excrete?
5 Why should you drink more, and eat more salt, in hot weather than in cold weather?

Topic 5 exercises

Transport in plants

1 Place a stick of celery complete with leaves in water which has been stained red or blue with ink. After about 30 minutes note the colour of the leaf veins. Cut across the stem and note where the stain appears inside the stem tissues. This indicates where the transport tubes are located. Slice the stem lengthwise to trace these tubes up and down the stem.

2 In late autumn look for leaf skeletons. These are formed when all but the leaf veins have rotted away. Use a magnifying glass to see how veins branch from the mid-rib (main vein) of a leaf.

The heart

3 Buy a heart from a butcher. Identify its outer parts from Figure 2 of Unit 5.3. Cut open the ventricles and main blood vessels and look for valves.

4 Open and close your hand fifty times. Could you go on doing this continuously for the rest of your life? Your heart muscles perform similar movements throughout your lifetime. What does this tell you about the difference between heart muscle and the muscles which move your hand?

The pulse

5 Find your pulse by placing the fingertips of one hand on your wrist, as shown in the photograph below. Count the number of pulses in one minute while you are sitting quietly. Count them again after doing some exercise. Why does the pulse rate change after exercise?

Blood

6 Some snake poisons destroy red blood cells. Why does this kill the victim?

7 A drop of blood was placed on a piece of glass. After one minute the tip of a needle was pulled through the blood. The needle came away from the blood trailing a long fine fibre. When does blood form these fibres? What is their function?

General

8 Match the following words with the statements below which describe them: diffusion, carbon dioxide, ventricle, white blood cells, atrium, haemoglobin, coronary artery, pocket valves, ureter, bladder, sweat glands.

   a) Some of them eat germs.

   b) The movement of dissolved substances from where they are plentiful to where they are scarce.

   c) They stop blood flowing back into the heart from the arteries.

   d) Stores urine.

   e) Transported from the body to the lungs mainly by the plasma.

   f) Has thick muscular walls.

   g) Help cool the body in hot weather.

   h) A chemical in red blood cells which enables them to transport oxygen.

   i) Transports urine from a kidney to the bladder.

   j) Has thin muscular walls.

   k) Carries the blood supply to the heart.

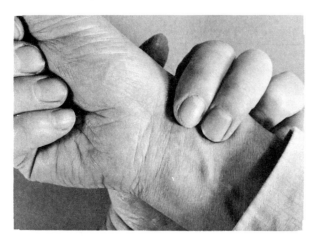

This photo shows how to feel your pulse

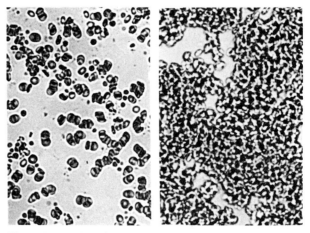

These photos show fresh blood from a pricked finger. Within a few seconds the red blood cells start to clump together, as shown on the left. In the photo on the right the cells are forming a clot.

Support and movement

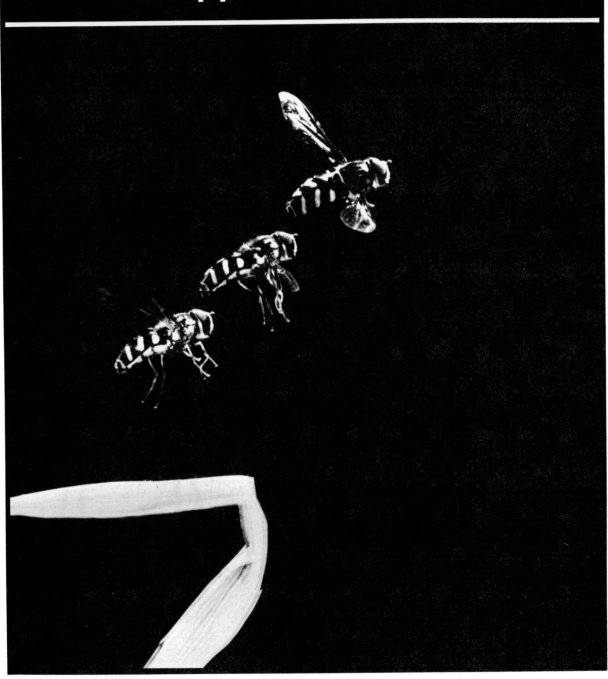

A hoverfly moves its legs and wings as it takes off

6.1 Types of support

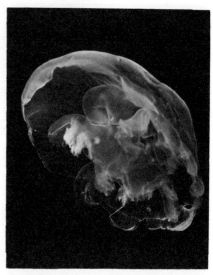

A jellyfish has no hard skeleton.

The amount of support a creature needs depends upon where and how it lives. Sea water supports jellyfish and sea weeds, but most land creatures would collapse into a floppy jelly without a rigid supporting skeleton of some kind.

Soft-bodied creatures

A jellyfish is helpless on land because it needs the support of the water in which it normally floats. Slugs, worms, and caterpillars, however, are almost as soft as jellyfish but can move about on land. This is possible because they are supported by water *inside* their bodies. Water fills their cells, and the spaces inside their bodies, inflating them like the air used to inflate a tyre. This enables them to withstand the pull of gravity. A caterpillar deflates like a punctured tyre if its skin is pricked, letting the supporting fluid drain out.

Young plant seedlings and the leaves and petals of plants are also supported by the pressure of water in their cells.

External skeletons (exoskeletons)

Arthropods, such as lobsters, crabs, and insects, are enclosed in an external skeleton, or **exoskeleton**, like a suit of armour (Fig. 1). An exoskeleton is made up of hard plates and tubes, with flexible joints where the body bends.

An exoskeleton supports the arthropod, and protects its soft inner parts from damage, dirt, and germs. It also stops the body drying up, and provides rigid material for the attachment of muscles.

Growth is very difficult if you live permanently inside a suit of armour. Arthropods overcome this problem by **moulting**. When they become too big for their exoskeleton they break it open, climb out of it, expand in size, and grow a completely new one. This happens several times as they grow to adult size.

Internal skeletons (endoskeletons)

Fish, amphibia, reptiles, birds, and mammals are vertebrates. Vertebrates are supported by an internal skeleton, or **endoskeleton**, made up of bones and gristle. An endoskeleton consists of at least a skull, and a backbone or **vertebral column** made up of small bones called **vertebrae**.

Land vertebrates usually also have a rib cage and limb bones, and shoulder and hip bones which hold the limbs in place (Figs. 2 and 3). Bones support the body, form a system of rods and levers to which muscles are attached, and contain tissue called **bone marrow** which produces blood cells. The skull protects the brain, eyes, inner ear, and nasal organs, and the ribs protect the heart, lungs, and main blood vessels.

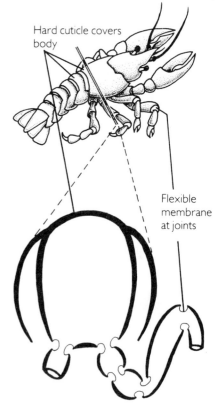

Hard cuticle covers body

Flexible membrane at joints

Fig. 1 A crayfish is supported by an external skeleton.

Exercises

1 What effect would gravity have on a jellyfish if it were taken out of water? Why doesn't gravity have this effect while the jellyfish is in water?

2 What supports soft-bodied land animals like slugs and earthworms?

3 Name three animals with an exoskeleton, and three with an endoskeleton. Choose animals not named in this Unit.

4 List the functions of an exoskeleton, and of an endoskeleton. Describe one thing an exoskeleton does which an endoskeleton does not do, and one thing that bones do which an exoskeleton does not do.

5 A tree trunk and branches are supported by wood. What supports the non-woody parts of a tree, like its leaves and flower petals?

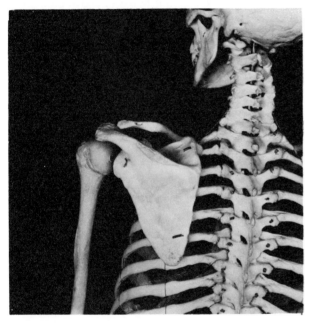

Back view of part of the human skeleton. Use Figure 3 to identify the bones.

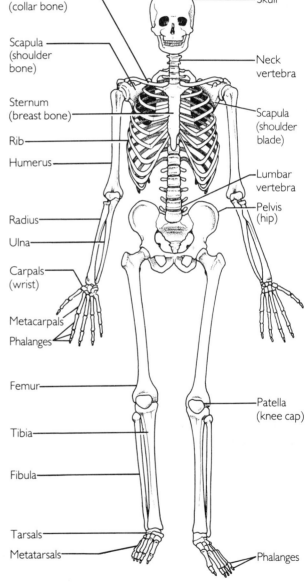

Fig. 3 The human skeleton

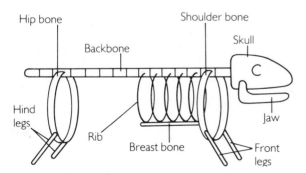

Fig. 2 General plan of an internal skeleton All vertebrates have a backbone (vertebral column) and a skull. Land vertebrates also have limbs. Shoulder and hip bones hold the limbs in place and take the weight of the body.

6.2 Bones, joints, and muscles

Together, the skeleton and muscles form an amazing piece of machinery. It is a machine which repairs its own broken parts, lubricates its own joints, works for years without wear, and can do jobs which require brute force like lifting weights, or which require delicate movements like writing a letter.

Bones

Bones consist of living cells embedded in a hard substance made mainly of calcium salts. Bones are held together at the joints by tough, flexible fibres called *ligaments* (Fig. 1).

Joints

The human skeleton has about seventy *moveable joints* (Fig. 1). The type of movement possible at a joint depends upon the shape of the bones where they rub together.

A *pivot joint* allows one bone to twist against another, like the joint beneath the skull which allows you to move your head from side-to-side (Fig. 2A). A *hinge joint*, such as the elbow, moves in only one direction, like the hinge of a door (Fig. 2B). A *ball-and-socket joint*, such as the hip, consists of a rounded head on one bone which fits into a cup-shaped socket on another (Fig. 2C and photo). This type of joint allows movement in several directions.

Muscles

A muscle is attached to a bone by a bunch of fibres called a *tendon* (Fig. 3B). Muscles cause movement of the body by contracting (getting shorter in length) so that they pull against the skeleton and make it bend at a joint.

Muscles can pull but cannot push. Therefore a muscle which bends a joint cannot straighten it again. Another muscle does this job by pulling in the opposite direction. Consequently, there are at least two muscles at each joint: a *flexor muscle* which bends (flexes) the joint, and an *extensor muscle* which straightens (extends) the joint (Fig. 3A and B).

Exercises

1 What are bones made of?

2 What is the difference between a ligament and a tendon?

3 Name *two* examples in the human skeleton of a hinge joint and of a ball-and-socket joint.

4 How is friction reduced where bones rub together at a joint?

5 Why does a joint require at least two muscles to move it?

6 What is the job of a flexor muscle, and of an extensor muscle?

7 Study Figure 3A and B. What are the main differences between these two joints? (Clue: look at the position of the muscles and skeleton in each.)

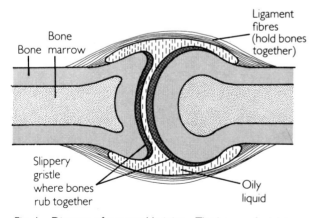

Fig. 1 Diagram of a moveable joint The bones of a joint are bound together by tough fibres called ligaments. Slippery gristle and oily liquid reduce friction where the bone ends rub together.

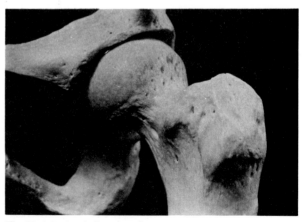

The hip joint is a ball-and-socket joint

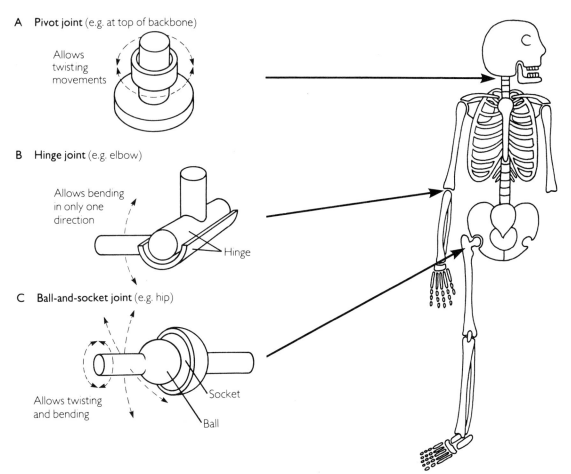

A **Pivot joint** (e.g. at top of backbone)

Allows twisting movements

B **Hinge joint** (e.g. elbow)

Allows bending in only one direction

Hinge

C **Ball-and-socket joint** (e.g. hip)

Allows twisting and bending

Socket

Ball

Fig. 2 Types of joint in the human skeleton

A **An insect leg joint**

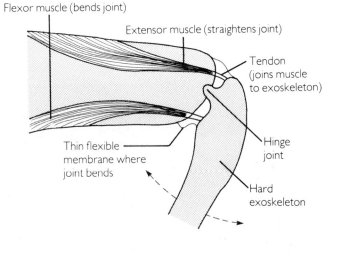

Flexor muscle (bends joint)

Extensor muscle (straightens joint)

Tendon (joins muscle to exoskeleton)

Hinge joint

Thin flexible membrane where joint bends

Hard exoskeleton

B **Human elbow joint**

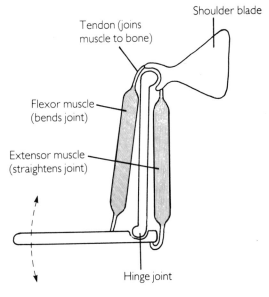

Shoulder blade

Tendon (joins muscle to bone)

Flexor muscle (bends joint)

Extensor muscle (straightens joint)

Hinge joint

Fig. 3 Arthropod and human muscle systems compared

6.3 Movement in water

The cheetah, which is the swiftest land animal, can run at 115 k.p.h. over short distances. Water is eight hundred times denser than air and yet the fastest fish, the Atlantic sailfish, can swim at speeds up to 110 k.p.h.

How fish swim

Almost all fish swim by sweeping their tail fin from side to side. A fish moves its tail fin by contracting powerful muscles on either side of its backbone (Fig. 1). These muscles contract first on one side of the backbone and then on the other, which bends the body and moves the tail from side to side (Fig. 2).

During swimming movements the tail fin presses backwards against the water and this pushes the fish forwards. It maintains a straight course by using its other fins as rudders and stabilizers.

Fish have a streamlined shape and many are made weightless in water by an air-filled space called a **swim bladder** inside their bodies.

Whales, porpoises, and dolphins

These are the only mammals which spend their whole lives in water. Unlike seals they do not even return to land for breeding. They swim using a horizontal tail fin which is swept up and down to push the body forwards and not from side to side like a fish tail (Figs. 3 and 4).

Although these animals live in water they have lungs and must come to the surface to breathe from time to time. They do this through nostrils on the top of the head (see Fig. 3 and photograph on p. 66). A whale can hold its breath for up to two hours, and can dive to a depth of 1500 m.

Exercises

1   What parts of its body does a fish use as it swims? Explain how these parts are used to move the fish through the water.

2   How is the shape of a fish similar to that of a dolphin? What is the advantage of this shape during swimming?

3   In what way is the swimming action of a fish different from a dolphin's? List some other ways in which dolphins and fish differ from each other.

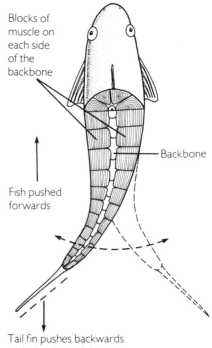

Blocks of muscle on each side of the backbone

Backbone

Fish pushed forwards

Tail fin pushes backwards against water as it is swept from side to side

Fig. 1 How fish swim   The top half of the fish is cut away to show muscles on each side of the backbone. These contract and relax alternately, causing the tail to sweep from side-to-side. The tail presses backwards against the water, which pushes the fish forwards.

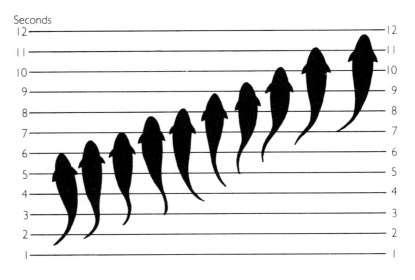

Fig. 2 A fish viewed from above as it swims
As the tail sweeps sideways it presses backwards against the water up till the moment when it reverses direction and repeats the same action.

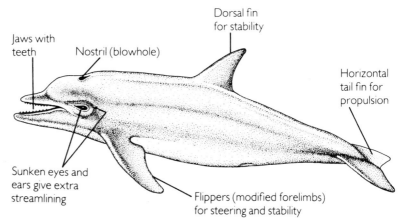

Jaws with teeth

Nostril (blowhole)

Dorsal fin for stability

Horizontal tail fin for propulsion

Sunken eyes and ears give extra streamlining

Flippers (modified forelimbs) for steering and stability

Fig. 3 A dolphin has a streamlined fish-like body. It swims by sweeping the tail fin up and down. Dolphins and whales have lungs and come to the surface to breathe through a nostril.

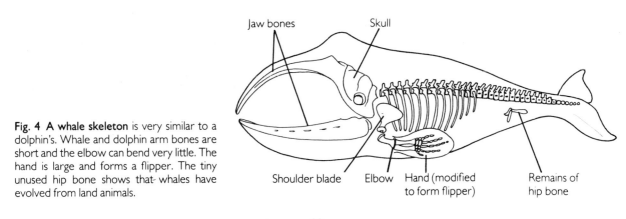

Jaw bones

Skull

Shoulder blade

Elbow

Hand (modified to form flipper)

Remains of hip bone

Fig. 4 A whale skeleton is very similar to a dolphin's. Whale and dolphin arm bones are short and the elbow can bend very little. The hand is large and forms a flipper. The tiny unused hip bone shows that whales have evolved from land animals.

6.4 Movement on land

Animals move from place to place in search of food. Herbivores can do this at a leisurely pace because plants don't run away. But when a herbivore is attacked by a fast-moving carnivore it must stand and fight or run for its life.

Sprawlers

Newts and lizards can be described as sprawlers because, except during very rapid movement, their bellies drag along the ground. Their limbs are only strong enough to lift their bodies off the ground for a few seconds (Figs. 1 and 2).

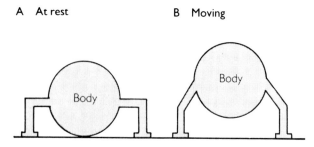

A At rest   B Moving

Fig. 1 Sprawlers only lift their bodies off the ground when moving quickly.

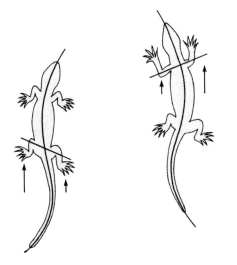

Fig. 2 A lizard moving The body is twisted into S-shaped waves, and the claws are used to drag the body along.

Walkers, runners, and jumpers

Most mammals have limbs powerful enough to lift their bellies off the ground during movement.

Cats A cat walks permanently on tiptoe (Fig. 3A). It can jump, climb, and run. Soft pads under its toes act as cushions allowing it to move very quietly when stalking prey.

Horses A horse also walks on tiptoe but, unlike a cat, a horse's foot consists of a very large toe protected by a hoof. This single toe corresponds to the middle toe of a five-toed animal. The long **cannon bone** in a horse's leg is equivalent to a bone from the palm of the human hand or sole of the human foot (Fig. 3B and C).

When walking a horse normally has three feet on the ground acting as a tripod which supports its weight while it moves the fourth foot. When it trots, a horse has two feet on the ground while it moves the two diagonally opposite feet.

Rabbits A rabbit moves by jumping, and can leap long distances to escape a predator. When at rest its hind legs are folded into a Z-shape and the whole foot rests on the ground (Fig. 4). When it is leaping, the hind legs are straightened by powerful muscles.

This lizard is moving slowly. It is dragging its belly along the ground.

Exercises

1 a) Why are newts and lizards described as sprawlers?
b) Why do sprawlers lift their bodies off the ground when moving very quickly?

2 Describe at least three differences between the feet of a cat and the feet of a horse.

3 Why does a cat need claws and the ability to walk quietly?

4 Why does a horse need the ability to run quickly?

5 What part of the human foot is a horse's cannon bone equivalent to?

6 Compare the way a horse walks with the way a lizard walks.

7 Compare the hind leg of a rabbit (Fig. 4) with the hind leg of a grasshopper (Unit 1.7), and with the hind leg of a frog (Unit 1.9). How are these legs similar, and how are they different?

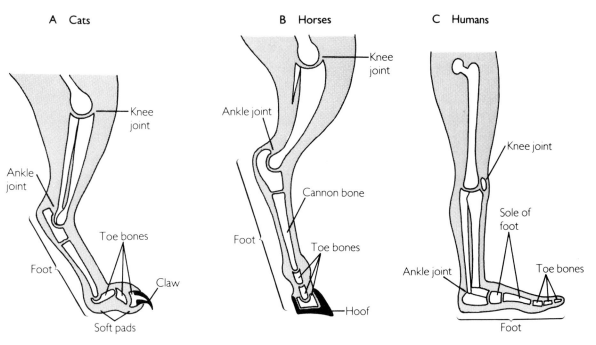

Fig. 3 Walkers and runners (hind legs, not drawn to scale)

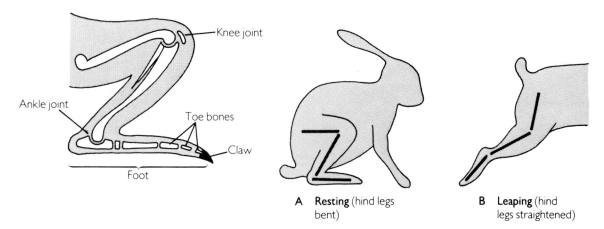

Fig. 4 A rabbit jumps and leaps by contracting powerful muscles which straighten its back legs.

6.5 Movement through air

Bats, birds, and some insects can defy the force of gravity and take to the air.

How birds fly

A bird's wings have two functions. They provide a force called lift which overcomes the force of gravity, and they flap round and round like propellers driving the bird forward through the air.

Lift   A bird's wing seen in cross-section has a curved shape called an *aerofoil* (Fig. 1B). As an aerofoil moves through the air its shape forces air to move faster across its upper surface than across its lower surface. Fast-moving air has a *lower* pressure than slow-moving air. Consequently a moving wing has a low pressure air stream above it and a high pressure air stream under it. The difference between these two pressures is an upward force—the lift (Fig. 1C). A bird stays in the air if lift equals or is greater than the weight of its body.

Wing movements   Wings do not simply move up and down. During each downstroke wing muscles pull the wing downwards and forwards. The front edge of the wing is lower than the rear edge so the wing pushes backwards against the air, which drives the bird forwards (Fig. 2A). During each upstroke the front edge of the wing is raised higher than the rear edge (Fig. 2B). Air now hits the underside of the wing and flips it upwards and backwards ready for the next downstroke.

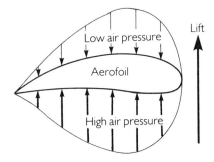

A   A bird's wing feathers

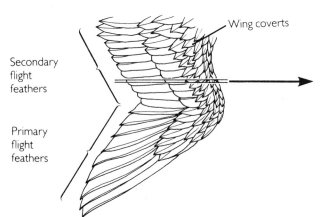

Secondary flight feathers

Primary flight feathers

Wing coverts

B   Cross-section of a wing showing its aerofoil shape

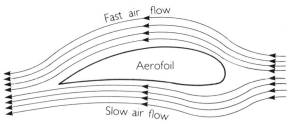

Fast air flow

Aerofoil

Slow air flow

C   Fast-moving air is at a lower pressure than slow-moving air. The difference between the two pressures equals lift

Low air pressure

Aerofoil

High air pressure

Lift

Fig. 1 How a bird's wing produces lift   During flight a wing has lower air pressure above it than underneath (**B** and **C**). The difference between these two pressures is the lift which keeps birds in the air.

Exercises

1 a) What is the technical name for the shape of a bird's wing seen in cross-section?
b) How does this shape produce the lifting force which keeps a bird in the air?
c) How does a bird drive itself forwards through the air?

2 a) Study the bones of the human arm illustrated in Figure 3B. How do these bones differ from those in a bird's and a bat's wing (Fig. 3A and C)?
b) How is a bat's wing different from a bird's wing?
c) Describe at least two ways in which a fly's wing (Fig. 3D) is different from a bird's.

A Downstroke   The front edge of the wing is lower than the rear edge, so the wing pushes backwards against the air. This drives the bird forwards.

B Upstroke   The front edge of the wing is above the rear edge so air hits the underside of the wing, flipping it upwards and backwards ready for the downstroke.

Fig. 2 Wing movements during flight

A   Bird's wing bones

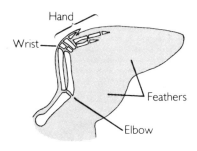

B   Human arm bones

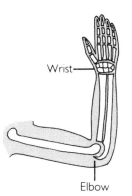

C   Bat's wing bones

D   Housefly's wing

Fig. 3 Compare these drawings   The bones in a bird's and bat's wings are similar to those in a human arm, but a fly's wing is completely different in structure.

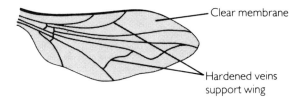

Topic 6 exercises

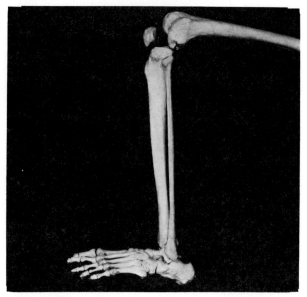

Bones of the human lower leg and foot

1   How many bones do you have? Copy the chart below into your notebook. Fill in the missing numbers by feeling bones through your skin. Figure 3 of Unit 6.1 will be useful. Don't forget to count *both* arms and legs.

Parts of the skeleton	Number of bones
Skull..	21
Lower jaw	
Backbone	33
Shoulder blades............................	
Collar bones.................................	
Breast bone	1
Ribs	
Hips..	2
Arms..	
Wrists	16
Hands	
Legs..	
Knee caps	
Ankles	14
Feet..	
Total	

2   What are the scientific names for the following bones: collar bone, backbone, shoulder blade, hip bone, breast bone, forearm bones?

3   a) Which of the creatures in Figure 1 has no hard skeleton, which has an exoskeleton, and which has an endoskeleton?
b) Can a snail shell be counted as an exoskeleton? What supports the parts of a snail's body which are not enclosed in the shell?
c) Both fish and sea anemones live in water. Why does a fish need a skeleton and not a sea anemone?

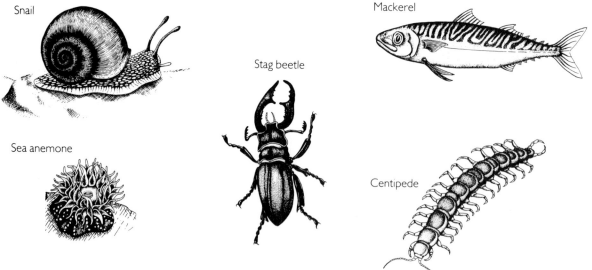

Snail

Sea anemone

Stag beetle

Mackerel

Centipede

Fig. 1   See exercise 3

Sensitivity and co-ordination

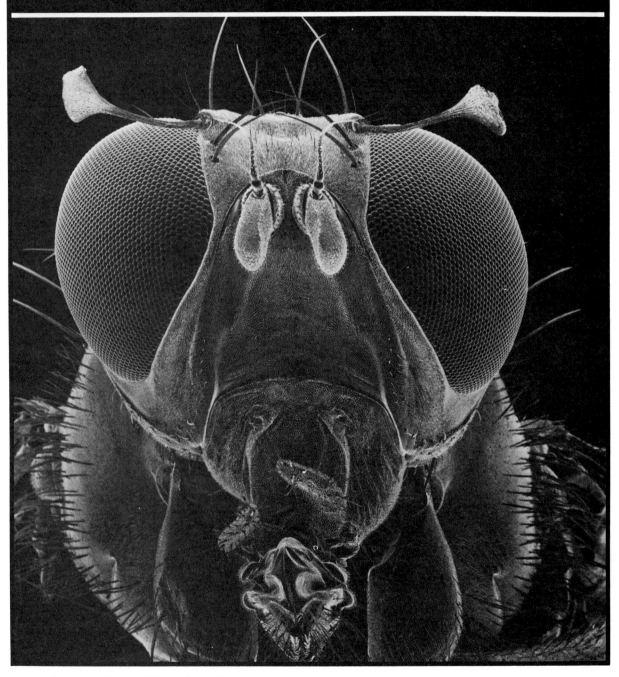

An insect's eyes contain many light-sensitive units

7.1   Plant senses

Plants cannot see, hear, or smell things but this does not mean they are insensitive to the world around them. They do respond to light, gravity, and water, and some respond to touch.

Light

Plants need light to make food by photosynthesis, and to make chlorophyll, the green substance in leaves which absorbs light energy. It is not surprising, therefore, that potted plants on a window sill bend over to face the light (Figs. I and 2), and that no matter which way up you plant a seed it will send its shoot straight upwards through the soil towards light (Fig. 3).

Gravity

Plant roots grow downwards in response to the pull of gravity. This ensures that roots always grow into the soil where they are most likely to find the water and minerals essential for growth. Plant shoots grow upwards in the opposite direction to the pull of gravity (Fig. 3).

Water

Normally roots find water by growing straight downwards, but if water is scarce in this direction but plentiful to one side, roots are capable of sensing the water and growing towards it, despite the pull of gravity (Fig. 4).

Touch

Insect-eating plants respond in various ways when touched by insects (see Unit 2.5).

If the tendrils of sweet peas or vines touch anything they respond by curling around it (Fig. 5). This is how they obtain the support they need to climb towards sunlight. The same thing happens if the seedlings of a dodder plant touch the stem of its host plant (described in Unit 3.6).

Exercises

I   Of what use to a plant is:
   a) its ability to grow towards light?
   b) the fact that its shoot grows in the opposite direction to the pull of gravity?
   c) the fact that its roots grow down in response to gravity?
2   Under what circumstances does a root 'ignore' the pull of gravity.
3   Why is a sense of touch important to climbing plants?
4   Answer the questions asked in Figures 2, 3, 4, and 5.

Sundew is sensitive to touch. When an insect gets stuck in the glue on a leaf, the tentacles bend over and trap the insect.

A   Plants grown out of doors grow straight up in response to light from above

Light

B   Plants grown on a window sill bend to face the light

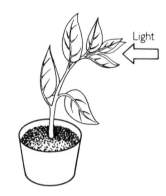

Light

Fig. I   Plants are sensitive to light

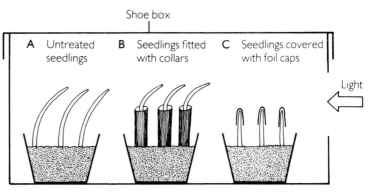

Fig. 2 Which part of a plant responds to light? Three sets of oat seedlings were grown in a shoe box which had a hole cut in one end. Set **A** was left uncovered. Set **B** was fitted with collars made of black paper which left the tips uncovered. The tips of set **C** were covered with aluminium foil. Set **A** and **B** seedlings bent towards the light. What can you conclude from this?

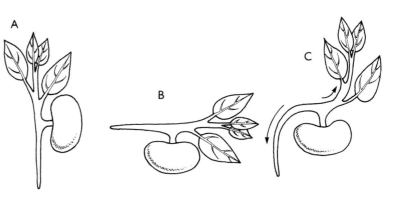

Fig.3 How do plants respond to gravity? A bean seedling was grown in an upright position (**A**). It was then turned onto its side (**B**) and placed in the dark. Its roots curved downwards and its shoot curved upwards (**C**). What can you conclude from this?

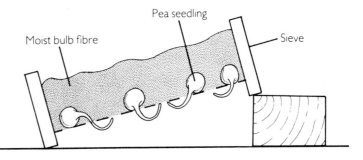

Fig. 4 How do roots respond to water? Some pea seedlings were grown in the bottom of a sieve covered with moist bulb fibre. At first their roots grew straight downwards through the sieve, then they curved upwards into the bulb fibre. What are the roots responding to: gravity or water? Why was the sieve tilted?

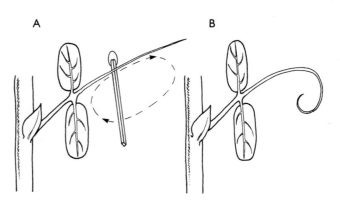

Fig. 5 How do pea tendrils respond to touch? A straight pea tendril was stroked with a match as shown in **A**. It curved as shown in **B**, while untouched tendrils on the same plant remained straight. What can you conclude from this?

101

7.2 The skin, tongue, and nose

Your skin is sensitive to four different types of stimulation and your nose and tongue allow you to detect countless different smells and flavours.

Skin

Human skin has nerve endings sensitive to touch, pressure, pain, and temperature (Fig. 1). These, and other sensory nerve endings, are called **receptors** because they 'receive' stimulation from the outside world.

Touch and pressure   Touch and pressure receptors are concentrated in the skin of the tongue and finger-tips. These receptors detect the texture of objects: whether they are rough, smooth, hard, or soft. Touch receptors are also attached to hair roots. If an object brushes against any hair the receptors are stimulated.

Pain   Pain receptors are more evenly distributed over the skin, and are also found inside the body in most tissues and organs. Pain acts as a warning signal. It tells the brain that something is wrong with the body.

Temperature   There are separate 'heat' and 'cold' receptors in the skin. These are used to detect changes in temperature. The finger-tips can detect temperature differences as small as 0.5 °C.

Nose

Smell receptors are called **olfactory organs**. They are sensitive to chemicals in the air (Fig. 2). But the chemicals must first dissolve in the film of moisture which covers the receptors.

Tongue

The taste receptors, or **taste buds**, in the tongue are also sensitive to chemicals. There are four types of taste bud: those sensitive to salt, sweet, sour, and bitter-tasting substances. The many different flavours of food and drink are identified according to how much they stimulate these four types of receptor. Groups of each type of receptor are concentrated in certain areas of the tongue (Fig. 3).

Certain tastes and smells are extremely unpleasant. These are usually produced by poisonous, decaying, or harmful substances. So taste and smell receptors protect you from harm in a similar way to pain receptors.

Exercises

1   If you were blindfolded and given a number of different objects to handle, what could you find out about them using only the sensitivity of your fingers?

2   A few people are born without a sense of pain. List some of the ways in which this disability puts their health, and their lives, at risk.

3   Cats and mice have long whiskers which grow out to approximately the width of their bodies. In what ways are these whiskers useful, and how would these animals be inconvenienced if their whiskers were cut off?

4   a) What are olfactory organs?
b) When you have a cold in the nose, the lining of your nasal passages is covered with a layer of mucus. Why does a cold reduce your sense of smell?

5   What are taste buds, and what types of taste are they sensitive to? How do you identify flavours?

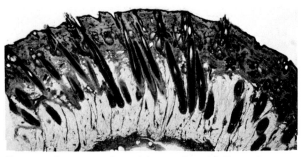

Blind people use touch receptors in their fingertips to read Braille

Compare this section through human skin with Figure 1

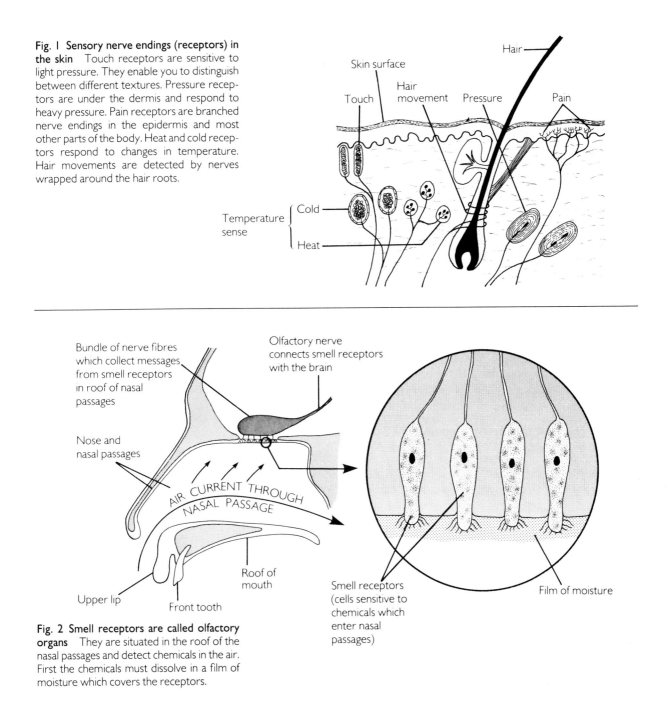

Fig. 1 Sensory nerve endings (receptors) in the skin Touch receptors are sensitive to light pressure. They enable you to distinguish between different textures. Pressure receptors are under the dermis and respond to heavy pressure. Pain receptors are branched nerve endings in the epidermis and most other parts of the body. Heat and cold receptors respond to changes in temperature. Hair movements are detected by nerves wrapped around the hair roots.

Skin surface

Hair

Touch

Hair movement

Pressure

Pain

Temperature sense { Cold Heat

Bundle of nerve fibres which collect messages from smell receptors in roof of nasal passages

Olfactory nerve connects smell receptors with the brain

Nose and nasal passages

AIR CURRENT THROUGH NASAL PASSAGE

Upper lip

Front tooth

Roof of mouth

Smell receptors (cells sensitive to chemicals which enter nasal passages)

Film of moisture

Fig. 2 Smell receptors are called olfactory organs They are situated in the roof of the nasal passages and detect chemicals in the air. First the chemicals must dissolve in a film of moisture which covers the receptors.

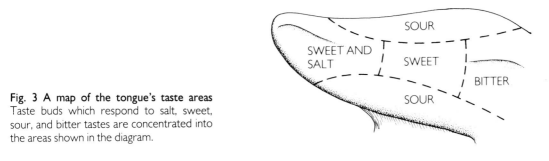

SOUR

SWEET AND SALT

SWEET

BITTER

SOUR

Fig. 3 A map of the tongue's taste areas Taste buds which respond to salt, sweet, sour, and bitter tastes are concentrated into the areas shown in the diagram.

7.3 Eyes

Many living things are sensitive to light, including plants and even some unicellular creatures. But a few types of animals are capable of vision. This means they have eyes which form a picture image of the world around them.

The human eye

An eye is like a camera. Both have lenses which focus an upside-down picture on a surface which is sensitive to light. In cameras the light-sensitive layer is the film, and in eyes it is a layer of nerve endings called the *retina*. Eyes and some cameras have an *iris diaphragm*. This is a device which controls the amount of light which falls on the film or retina (Figs. I and 3).

Unlike a camera, an eye does not form a permanent picture. The retina changes light into nerve impulses which pass down the *optic nerve* to the brain. The brain changes these impulses into a three-dimensional impression of the outside world.

Cornea This is the transparent window at the front of an eyeball (Figs. I and 3A).

Iris This is the coloured part of the eye. It has a hole at its centre called the *pupil* (Figs. I and 3A). In dim light, muscles in the iris make the pupil larger which lets more light into the eye. Other iris muscles make the pupil smaller in bright light.

Lens The lens of an eye is not hard like the glass lens of a camera. Muscles stretch it during focusing. The lens is held in place by *suspensory ligaments* (Fig. 3A).

Insect eyes

Insects, and many other arthropods, have both *simple eyes* and *compound eyes* (Fig. 4A). Compound eyes are made up of many separate visual units (Fig. 4B). Dragonflies have 28 000 of these units in each compound eye. Each unit consists of a lens and light-sensitive cells (Fig. 4C). These cells send impulses to the insect's brain where they are put together to make a visual impression which must be something like a picture made up of mosaic tiles. Simple eyes are made up of one visual unit.

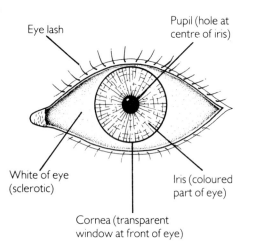

Fig. I **Front view of the human eye**

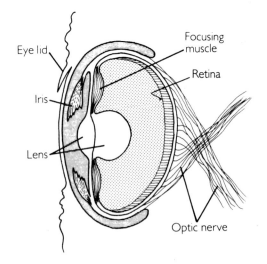

Fig. 2 **Eye of an octopus** This is the most advanced and efficient eye of any invertebrate animal. Compare it with the human eye (Fig. 3).

Exercises

1 Which parts of the human eye fit the following descriptions?

a) The coloured part of the eye.

b) Gets bigger in dim light, and gets smaller in bright light.

c) A window at the front of the eyeball.

d) Stretched during focusing.

e) A hole in the iris.

f) Hold the lens in place.

g) A layer of light-sensitive cells.

h) Nerve connecting an eye with the brain.

2 List the ways in which a camera and a human eye are similar. In what ways are they different?

3 What is the function of the iris?

4 The inside of a camera is painted matt black. Beneath the retina of the eye is a layer of black substance called the choroid (Fig. 3A and B). The black paint in a camera and the choroid of the eye share the same function. What is this function?

5 In what ways are an insect's compound eyes and human eyes different? Are they similar in any way?

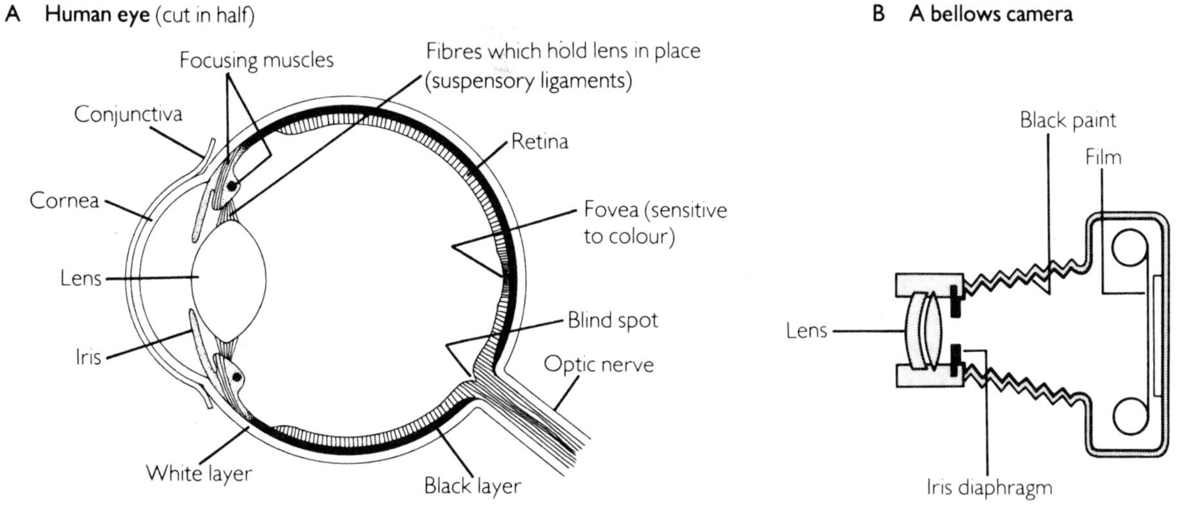

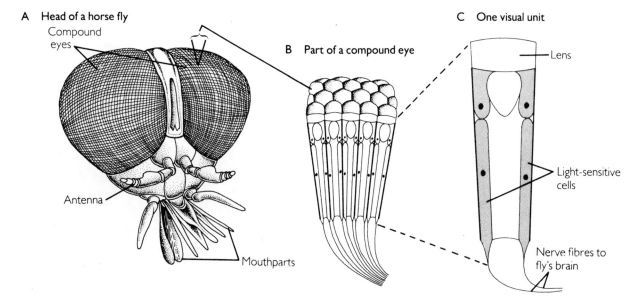

A   Human eye (cut in half)

- Focusing muscles
- Fibres which hold lens in place (suspensory ligaments)
- Conjunctiva
- Retina
- Cornea
- Fovea (sensitive to colour)
- Lens
- Blind spot
- Iris
- Optic nerve
- White layer
- Black layer

B   A bellows camera

- Black paint
- Film
- Lens
- Iris diaphragm

Fig. 3 Compare the human eye with a camera

A   Head of a horse fly

- Compound eyes
- Antenna
- Mouthparts

B   Part of a compound eye

C   One visual unit

- Lens
- Light-sensitive cells
- Nerve fibres to fly's brain

Fig. 4 Insect eyes

7.4 Vision

Human eyes are capable of seeing an object as small as one-tenth of a millimetre from a distance of 25 cm, and can see 10 000 000 different shades of colour.

Three-dimensional (3-D) vision

3-D vision is best developed in animals whose eyes face forwards, such as humans and apes. This is because 3-D vision depends upon both eyes looking at the same object. Human eyes are about 6 cm apart, and when you look at something each eye sees a slightly different view (Fig. 1). These two views are put together in the brain to form a 3-D impression.

Focusing

Unlike a camera the lens of the eye is not moved back and forth to change focus and form a clear image on the retina (Fig. 2C and D). The eye is focused by two sets of muscles which change the shape of the lens. To focus on an object more than 10 metres away a set of *radial muscles* in the eye pull against the suspensory ligaments. This stretches the lens into a thin shape (Fig. 2A).

To focus on a nearer object the radial muscles relax and *circular muscles* contract, pulling the walls of the eyeball inwards. This slackens the suspensory ligaments and allows the lens to bulge outwards into a fatter shape (Fig. 2B).

Colour vision

The retina is not the same all over. There is a small circular area opposite the lens called the *fovea* which is sensitive to colour (Fig. 3, p. 105). The fovea only works in bright light, which explains why colours are less easy to see towards evening. The remainder of the retina is not sensitive to colour, but works even in dim light.

Exercises

1 Why do rabbits have poorer 3-D vision than monkeys?

2 a) What shape is the lens in a human eye when the eye is focused on a near object? and on a distant one?
b) What part do circular and radial muscles play in changing the shape of the lens during focusing?
c) Compare focusing a camera with focusing in the eye.

3 Why should you always look at the colour of a garment in bright light before buying it?

4 How wide is your angle of vision? To find out hold your arms straight out in front of you with your thumbs upwards. Move your arms sideways away from each other. Keep staring straight ahead but notice when the thumbs are still just visible out of the corner of each eye. Have a partner estimate the angle of your arms. How does it compare with a horse's field of vision (Fig. 4)?

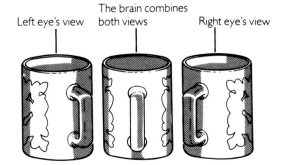

Left eye's view   The brain combines both views   Right eye's view

Fig. 1 Three-dimensional vision depends upon two eyes looking at the same object. Human eyes are 6.3 cm apart and so each sees a slightly different view (exaggerated in the two outer drawings). The brain combines these two views to make a 3-D impression of the object (centre drawing).

The out-of-focus photo on the left shows how an object looks to someone with poor vision. The photo on the right shows normal vision or poor vision corrected by spectacles.

A Eye focused on a distant object The radial muscles contract and pull against the suspensory ligaments. This stretches the lens into a flattened shape.

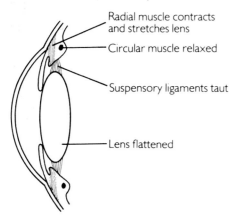

Radial muscle contracts and stretches lens

Circular muscle relaxed

Suspensory ligaments taut

Lens flattened

B Eye focused on a near object Circular muscles contract. This releases tension on the suspensory ligaments and the lens becomes more rounded in shape.

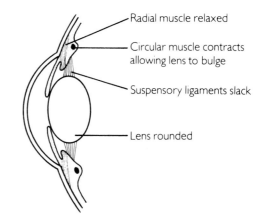

Radial muscle relaxed

Circular muscle contracts allowing lens to bulge

Suspensory ligaments slack

Lens rounded

C Camera focused on a distant object
Lens moved backwards

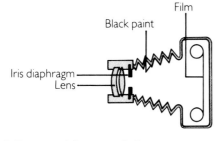

Film

Black paint

Iris diaphragm
Lens

D Camera focused on a near object
Lens moved forwards

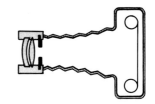

Fig. 2 Focusing of the eye, and of a camera

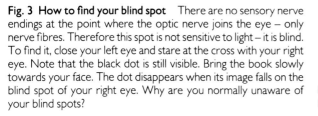

Fig. 3 How to find your blind spot There are no sensory nerve endings at the point where the optic nerve joins the eye – only nerve fibres. Therefore this spot is not sensitive to light – it is blind. To find it, close your left eye and stare at the cross with your right eye. Note that the black dot is still visible. Bring the book slowly towards your face. The dot disappears when its image falls on the blind spot of your right eye. Why are you normally unaware of your blind spots?

320°

Fig. 4 A horse's angle of vision is 320° What is yours? Find out by doing exercise 4.

7.5 Balance and hearing

Your ears transform vibrations in the air (sound waves) into nerve impulses. These pass to the brain where they give you the sensation called sound. Your ears also contain organs which help you keep your balance.

Balance

Three types of sensation help you to keep your balance. Messages from your eyes tell you if you are upright or tilted; messages from your muscles tell you, without your looking, about the position of your limbs and trunk; and the brain combines both of these messages with a third set from your *semi-circular canals* (Figs. 1 and 2).

Semi-circular canals are curved tubes filled with a liquid which moves when you move. Nerve endings in the canal walls carry information about the fluid's movement to the brain, and the brain uses this information to send instructions to the muscles you use in keeping your balance (Fig. 1).

Hearing

Sound waves are collected by the funnel-shaped *pinna* of each ear, and pass down a short tube to the *ear drum* (Figs. 2 and 3). The ear drum is a thin sheet of skin, and sound waves make it vibrate.

Behind the ear drum there is an air-filled space which contains a chain of three tiny bones called the *ear ossicles*. These bones connect the ear drum with another sheet of skin called the *oval window*.

When the ear drum vibrates, the ear ossicles move against each other in such a way that they lever the oval window in and out. This causes vibrations to pass along a tube called the *cochlea*.

The cochlea is coiled like the shell of a snail and is filled with liquid (Fig. 2). As vibrations move along the cochlea they cause a layer of nerve endings to shake up and down (Fig. 3). This stimulates the nerve endings into sending nerve impulses along the *auditory nerve* to the brain, where they are interpreted as sounds.

In other words, it is the cochlea which transforms sound waves into the nerve impulses which give the sensation of sound.

A **Position of the semi-circular canals in the head**
(Canals drawn larger than life)

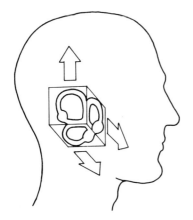

Fig. 1 The semi-circular canals

B **Functions of the semi-circular canals**

Vertical canals (stimulated by nodding and sideways head movements)

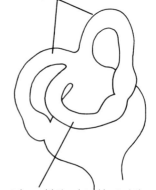

Horizontal canal (stimulated by twisting movement, such as when you shake your head to say 'No')

Exercises

1   Why would it be dangerous for a person with defective semi-circular canals to ride a bicycle?

2   Explain how the ears transform sound waves into nerve impulses.

3   Match the following words with the descriptions below: cochlea, pinna, ear drum, ear ossicles, semi-circular canals, auditory nerve, oval window.

a) Funnels sound waves towards the ear drum.

b) Coiled tube which changes sound waves into nerve impulses.

c) A chain of bones.

d) When sound waves make it vibrate it moves the ear ossicles.

e) Organs of balance.

f) Levered in and out by the ear ossicles.

g) Carries nerve impulses from the cochlea to the brain.

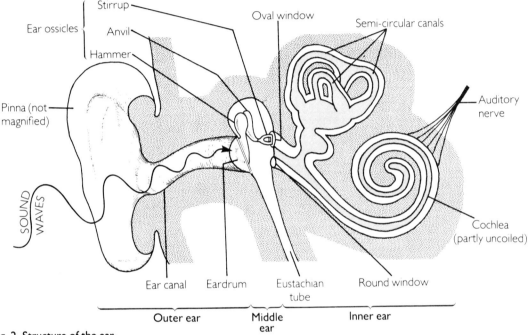

Fig. 2 Structure of the ear

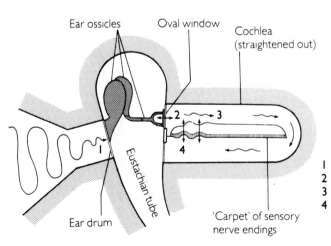

Fig. 3 Diagram of how the ear works (cochlea straightened out)

1   Sound waves cause the ear drum to vibrate.
2   This causes the ossicles to move the oval window in and out.
3   This causes vibrations to move through fluid in the cochlea.
4   This causes sensory nerve endings in the cochlea to vibrate up and down, which stimulates them to send nerve impulses to the brain where they are interpreted as sounds.

7.6 Nervous systems

The cells, tissues, and organs of humans and other animals do not work independently of each other, but are co-ordinated. This means they work together, serving the body as a whole. Muscles are co-ordinated by the nervous system.

What is a nervous system?

All nervous systems are made up of cells whose job is to conduct signals called **nerve impulses** to and from all parts of the body. Like all other cells a nerve cell consists of a nucleus and cytoplasm, but unlike other cells the cytoplasm extends outwards, forming **nerve fibres**. In humans, nerve fibres can be up to 1 metre long. Nerves are bundles of nerve fibres (Fig. 2C).

Types of nervous system

Plants and single-celled creatures like *Paramecium* do not have a nervous system. But *Paramecium* does have **nerve fibrils** which co-ordinate movements of its cilia (Fig. 1A).

Hydras have proper nerve cells connected together to form a **nerve net** (Fig. 1B). When any part of a *Hydra* is stimulated impulses spread slowly from this point out in all directions through the nerve net. Only a strong stimulus affects all of a *Hydra*'s body. A nerve net provides only a very simple level of co-ordination.

An insect has a more complex nervous system than a *Hydra* (Fig. 1C). Most of an insect's nerve cells are clustered into groups called **ganglia** (the singular is ganglion). The largest ganglion is in the head region, and a chain of smaller ganglia extend along the body. These ganglia are connected by one or two **nerve cords**.

Cells in the ganglia send out nerve fibres into each segment of the insect's body. Ganglia receive impulses from sense organs and direct them to other parts of the body, particularly the muscles. This system gives insects a high level of co-ordination.

Humans and all other vertebrates have a single tubular nerve cord, the **spinal cord**. The brain is an enlarged and very complex extension of the spinal cord. Together, the brain and spinal cord are called the **central nervous system**.

The central nervous system has many nerves branching from it (Fig. 2A, B, and C). The human brain not only co-ordinates many activities in the body, it is also capable of remembering and thinking. A complex nervous system allows an animal to behave in a complex way.

Insects have much more complicated behaviour patterns than *Hydras*, and humans have the most highly developed nervous system and behaviour of any animal.

A Paramecium No nervous system. Only nerve fibrils

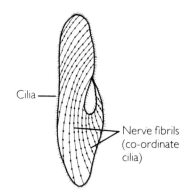

B Hydra Network of interconnected nerve cells

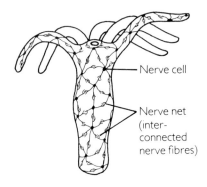

C Insects Ganglia connected by nerve cords

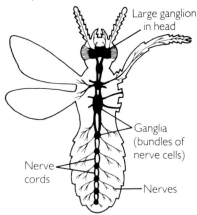

Fig. 1 Types of nervous system

Exercises

1 What is co-ordination and why is it important?

2 What is the function of a nerve cell, and how do nerve cells differ from all other cells?

3 What is the difference between a nerve and a nerve fibre?

4 Why is it correct to say that *Paramecium* has no nervous system even though it has structures capable of coordinating movements of its cilia?

5 In what ways is an insect's nervous system more complex than that of a *Hydra*?

6 a) What are the parts of the human central nervous system?
   b) What are the advantages of having a complex nervous system?

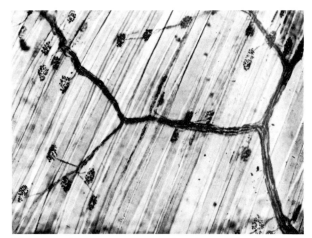

This photo shows motor nerve endings in a muscle

A   The human nervous system

B   The brain (and part of the spinal cord)

C   A nerve (made up of nerve fibres)

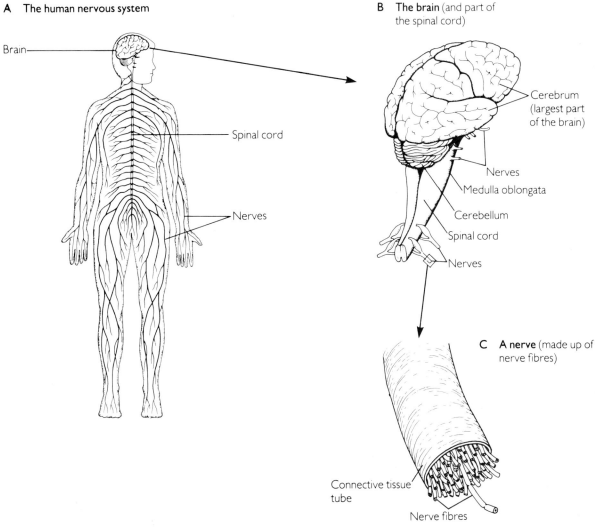

Fig. 2 **The human nervous system** is made up of nerves, a spinal cord, and a brain.

7.7 Reflexes and the brain

The human brain weighs 1.5 kg, but many animals have much heavier brains. A whale's brain weighs 9 kg, an elephant's 4.5 kg. When brain weight is compared with body weight, bird brains are proportionately bigger than ours. However, the human brain is more highly developed than that of any other creature.

Nerve cells

Nerve cells conduct impulses in only *one* direction. Impulses travel from sense organs into the central nervous system (brain and spinal cord) along *sensory nerve cells* (Fig. 1A). Impulses then travel along connecting nerve cells to *motor nerve cells* which conduct them to muscles (Fig. 1B). A *reflex action* is a simple example of nerve cells in action.

Reflexes

During a reflex action you respond to something automatically without having to think about it. If, for example, you accidentally touch something sharp you quickly pull your hand away without stopping to think what to do (Fig. 2). This reflex protects you from injury. Other reflexes are mentioned in Exercise 2.

The brain

Your brain contains thousands of millions of nerve cells and each of these is connected with a thousand or more other nerve cells making an immensely complex network of cells and nerve fibres.

The largest part of the brain is the *cerebrum* (Fig. 3). It receives information from all the sense organs and uses it to control behaviour. It is also the part of the brain you use to remember things, to solve problems, and to make decisions.

Below the cerebrum are the *cerebellum* and *medulla oblongata*, whose functions are listed in Figure 3.

Exercises

1 a) What is a reflex action?
   b) Describe the events which would occur in your nervous system if you accidentally touched something hot.

2 What reflex actions occur in response to the following types of stimulation:
   a) when dust blows into your eyes (2 reflexes)?
   b) when a bright light shines in your eyes (2 reflexes)?
   c) when you move from a warm room to a very cold one?
   d) when food accidentally enters your wind-pipe?

3 Of what use to the body is each of the reflexes described in the last question?

4 List the main functions of the cerebrum, cerebellum, and medulla oblongata.

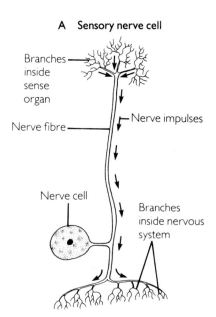

A   Sensory nerve cell

Branches inside sense organ

Nerve fibre

Nerve impulses

Nerve cell

Branches inside nervous system

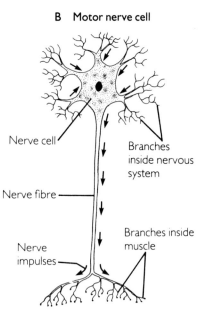

B   Motor nerve cell

Nerve cell

Branches inside nervous system

Nerve fibre

Nerve impulses

Branches inside muscle

Fig. 1 Nerve cells

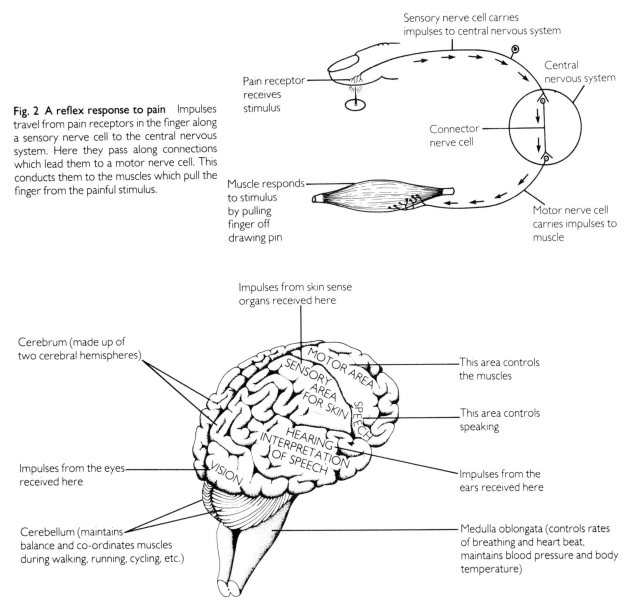

Fig. 2 A reflex response to pain Impulses travel from pain receptors in the finger along a sensory nerve cell to the central nervous system. Here they pass along connections which lead them to a motor nerve cell. This conducts them to the muscles which pull the finger from the painful stimulus.

Sensory nerve cell carries impulses to central nervous system

Pain receptor receives stimulus

Central nervous system

Connector nerve cell

Muscle responds to stimulus by pulling finger off drawing pin

Motor nerve cell carries impulses to muscle

Impulses from skin sense organs received here

Cerebrum (made up of two cerebral hemispheres)

MOTOR AREA

SENSORY AREA FOR SKIN

SPEECH

HEARING INTERPRETATION OF SPEECH

VISION

This area controls the muscles

This area controls speaking

Impulses from the eyes received here

Impulses from the ears received here

Cerebellum (maintains balance and co-ordinates muscles during walking, running, cycling, etc.)

Medulla oblongata (controls rates of breathing and heart beat, maintains blood pressure and body temperature)

Fig. 3 Parts of the human brain

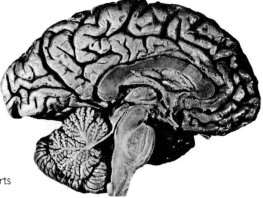

Left half of a human brain. Identify the parts from Figure 3.

Topic 7 exercises

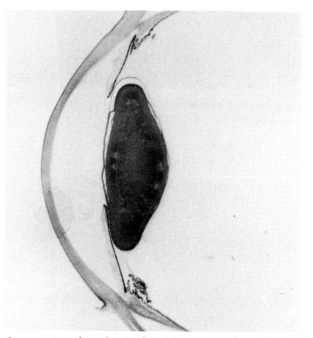

Cross-section of eye showing lens, iris, cornea, and pupil. Look at Fig. 3 on p. 105 and identify the parts.

Skin

1   A student was blindfolded and the skin of her finger-tips, palm, and arm was touched many times with two needle points. Sometimes the needle points were 0.5 cm apart, and sometimes they were 1.0 cm apart. Each time the needles touched, the student was asked if she thought one or two points were being used. The table below shows the number of times she was **correct** when she said two points were touching her.

Needle points	Arm	Palm	Finger-tips
1.0 cm apart	10	17	25
0.5 cm apart	8	12	14

What do these results tell you about the distance apart of touch sense organs in the skin of the finger-tips, palm, and arms?

Vision

2   a) Arrange two pencils on a desk top in the positions A and B shown in Figure 1. Sit so that your eyes are level with the desk surface. Close one eye and with the other look across the desk halfway between the two pencil points.
b) Using *one* hand only try to arrange the two pencils so that their points are opposite and about 2 cm apart (to positions C and D in Figure 1).
c) Try again using both eyes.
What do your results tell you about judging distances using one and two eyes?

Hearing

3   What can horses and rabbits do with their ears (pinnae) which we cannot do with ours? Of what use is this ability?

4   Find the Eustachian tube on Figure 2 of Unit 7.5. These tubes connect the air-filled space behind each ear drum with the back of the mouth. When you swallow the Eustachian tubes open, letting air in or out of these spaces so that air pressure is always the same on each side of the ear drums.
a) The Eustachian tubes often become blocked when you have a cold. When this happens air in these tubes is absorbed into blood vessels lining their walls. Why does this cause deafness?
b) Some airline companies offer passengers sweets to eat as the aircraft takes off. This helps prevent deafness. What causes the deafness, and why does eating a sweet help prevent it?

Nervous system

5   How long does it take for nerve impulses to travel from a pressure sense organ in one hand to a muscle in the other hand? Find out in the following way:
a) The class must form a line holding hands. The teacher should be at one end of the line holding a stop watch hidden from view, and the student at the other end of the line should be ready to bang a desk top with his free hand.
b) Without any warning the teacher starts the stop watch and at the same moment squeezes the hand of the first student. This pupil squeezes the next student's hand as quickly as possible and so on down the line. When the last student's hand is squeezed he bangs the desk. This is the signal for the teacher to stop the watch.
c) The time it takes for the signal to reach the end of the line divided by the number of students gives the approximate speed at which impulses can travel through one student's body.

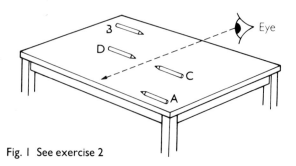

Fig. 1 See exercise 2

Reproduction and heredity

These millipedes are mating

8.1 Reproduction: asexual and sexual

Nothing lives for ever. Yet life continues, because living things have the ability to create new life by reproducing. New life is created in two ways: by asexual reproduction and by sexual reproduction.

Asexual reproduction

This type of reproduction involves only *one* parent. It occurs mainly in organisms whose bodies have a simple structure.

Amoeba   Unicellular organisms like *Amoeba* can reproduce in the simplest possible way: they split into two parts (Fig. 1).

Pin mould   Pin mould is given this name because of the shape of its reproductive organs (Fig. 2). This mould reproduces asexually by making thousands of microscopic cells called **spores**. Spores are made inside a bubble-like **spore case** which grows on top of a stalk. When a spore case is full of spores it bursts open, releasing spores which are distributed by wind and by insects. If a spore lands on a suitable food it splits open and forms a new mould colony.

Hydra   A *Hydra* reproduces asexually during the summer. It does so by a method called **budding**—the young hydras grow as bud-like growths on their parent (Fig. 3 and photograph at top of opposite page). Eventually a bud forms tentacles, separates from its parent, and floats away. After floating for a day or two it settles down to feed and grow to adult size. A *Hydra* is also capable of sexual reproduction.

Sexual reproduction

This type of reproduction involves *two* parents. Parents produce **sex cells** from **sex organs**. A male sex cell joins with a female sex cell. This event is called **fertilization** and results in a single cell called a **zygote**. The zygote grows into a new organism.

Hydra   A *Hydra* reproduces sexually in the autumn. Most hydras are **hermaphrodite**, which means an individual has both male and female sex organs. But these organs never develop at the same time on the same animals.

A male sex organ called a **testis** produces hundreds of sex cells called **sperms**. Sperms have tails which they use to swim through water until they find another *Hydra* with a ripe female sex organ (Fig. 4A). A female sex organ is called an **ovary**. It develops one large sex cell called an **ovum**. Only one sperm joins with the ovum. Together the sperm and ovum form a zygote which develops into a ball of cells (Fig. 4B). This ball forms a protective wall around itself, drops from its parent, and in the spring grows into a new *Hydra*.

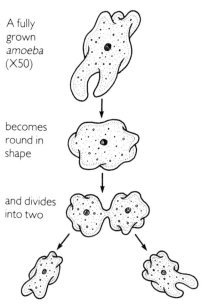

A fully grown *amoeba* (X50)

becomes round in shape

and divides into two

Fig. 1 Amoeba reproduces asexually by dividing into two

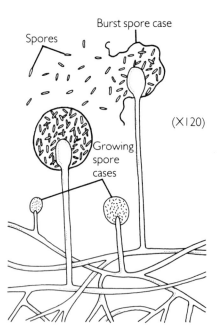

Spores

Burst spore case

(X120)

Growing spore cases

Fig. 2 Pin mould reproduces asexually by producing spores   If these land on suitable food they form a new mould colony.

Exercises

1 What is the main difference between asexual and sexual reproduction?

2 An *amoeba* can reproduce by splitting into two parts. Why would this be an impossible type of reproduction for humans to perform?

3 A pin mould colony as big as a 10p coin produces thousands of spores a day. What are spores for? Why are they produced in such large numbers?

4 a) How does a *Hydra* reproduce asexually?
b) Describe how a *Hydra* reproduces sexually.
c) Why does asexual reproduction by a *Hydra* occur only in summer, and sexual reproduction occur only in autumn?

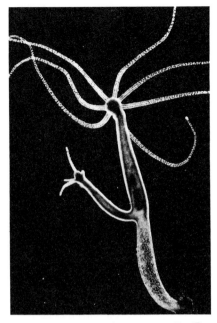

This Hydra is reproducing asexually. The bud will break off to form a new Hydra

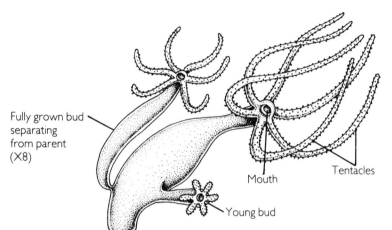

Fully grown bud separating from parent (X8)

Mouth

Tentacles

Young bud

Fig. 3 Asexual reproduction in a Hydra

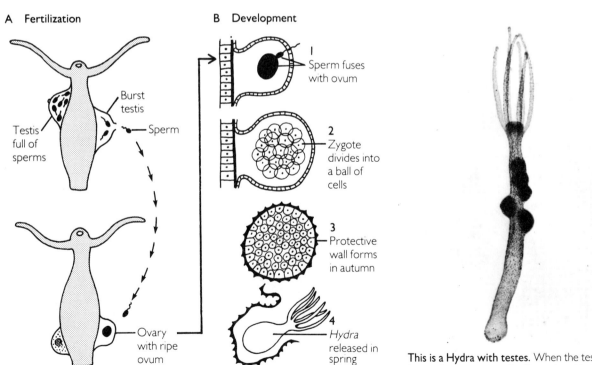

A   Fertilization

Testis full of sperms

Burst testis

Sperm

Ovary with ripe ovum

B   Development

1 Sperm fuses with ovum

2 Zygote divides into a ball of cells

3 Protective wall forms in autumn

4 *Hydra* released in spring

Fig. 4 Sexual reproduction in a Hydra

This is a Hydra with testes. When the testes burst, the sperm can fertilize other Hydras with ripe ova

8.2 Flowers

Not all flowers are coloured and scented. Grasses have unscented green flowers, and the largest flowers in the world are white with muddy brown spots, measure 90 cm across, and belong to the foul-smelling stinking corpse lily of South-East Asia.

Parts of a flower

Flowers have the following parts.

Carpels These are the female reproductive organs and are found at the centre of a flower. Buttercups have very simple carpels (Fig. 1). Each has a hollow base, the **ovary**, which contains a single **ovule** (Fig. 1B). An ovule contains the female sex cell. The ovary eventually becomes a **fruit**, protecting the ovule which becomes a **seed**. Above the ovary there is a narrow **style** which ends in a **stigma**. Pollen grains stick to the stigma during pollination.

Buttercups have several separate carpels, but in many flowers there are a number of carpels fused together.

Stamens These are the male reproductive organs. They are arranged in a ring around the carpels. A stamen consists of a stalk bearing an **anther** (Fig. 1C). Each anther is made up of four **pollen sacs** in which pollen grains grow (Fig. 1D). Pollen grains contain the male sex cells.

Petals In most flowers the carpels and stamens are surrounded by a ring of petals. Some flowers have coloured and scented petals with a **nectary** at the base which produces sugary **nectar** (Fig. 1).

Petals of this type attract insects which come to drink the nectar. As insects do this they transfer pollen from flower to flower.

Sepals Many flowers have an outer ring of sepals. These often look like small green leaves. Their function is to enclose and protect the flower when it is in the bud stage of development.

A Buttercup flower cut in half

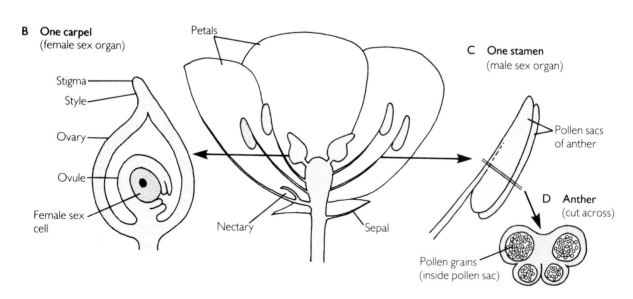

B One carpel (female sex organ)

Stigma
Style
Ovary
Ovule
Female sex cell

Petals

Nectary
Sepal

C One stamen (male sex organ)

Pollen sacs of anther

D Anther (cut across)

Pollen grains (inside pollen sac)

Fig. 1 Parts of a flower   (Based on the buttercup)

Exercises

Name the parts of a flower described by each of the following statements:

1 Male reproductive organs.
2 Often coloured and scented.
3 Female reproductive organs.
4 Pollen grains stick to it during pollination.
5 Contains the female sex cell.
6 Contains male sex cells.
7 Eventually becomes a fruit.
8 Produces a sugary fluid which attracts insects.
9 Eventually becomes a seed.
10 Form the outer covering of a flower bud.
11 Contains pollen grains.

A Whole sweet pea flower

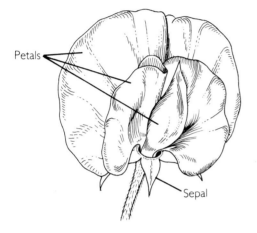

B Sweet pea flower cut in half

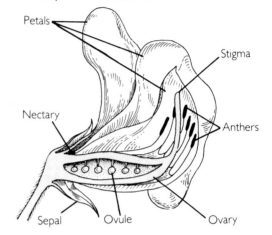

Fig. 2 The sweet pea flower

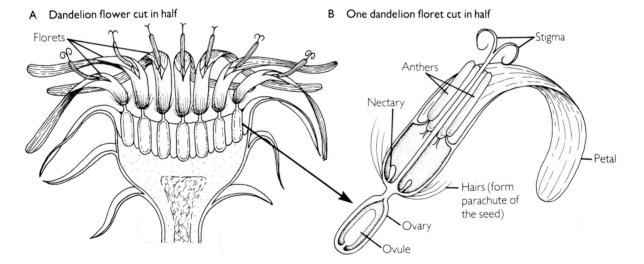

Fig. 3 A dandelion flower is really a collection of small separate flowers called florets. Each floret has its own set of reproductive organs.

119

8.3   Pollination and fertilization

Magnified pollen grains

A flowering plant's male sex cells are its pollen grains and its female sex cells are its ovules. But neither type of sex cell is capable of movement. So how are the two brought together for fertilization? The answer is that plants rely on two pollen delivery services.

Pollination

Pollination is the transfer of pollen from anthers to stigmas. This must occur before fertilization can take place. *Self-pollination* is the transfer of pollen within the same flower, or between flowers on the same plant, and *cross-pollination* is the transfer of pollen between flowers on different plants of the same type.

Wind-pollinated flowers   Grasses and stinging nettles rely on wind to deliver pollen from flower to flower. But the chances of a tiny pollen grain blowing straight from the anthers of one flower to the stigmas of another even a short distance away are very small indeed.

However, wind-pollinated plants have ways of overcoming these odds. First, they produce vast quantities of light-weight pollen which floats on the slightest breeze. Second, their flowers have spreading feathery stigmas which make a large target for pollen grains to hit (Fig. 1B). Third, flowers and anthers are on long stalks and sometimes develop before the leaves so they are well exposed to wind.

Insect-pollinated flowers   Buttercups, peas, dandelions, and dead nettles are examples of the enormous number of plants which make use of an extremely efficient pollen delivery service: insects.

Insects do not carry pollen from flower to flower free of charge. They are rewarded for their service in two ways. First, they can eat some of the pollen itself and second, many flowers produce nectar. Bees and butterflies find this sweet liquid so attractive that they spend almost all their lives in search of it.

Plants advertise their pollen and nectar with brightly coloured flowers and scents which insects find attractive. Often the brightest colours mark a path through the flower to where nectar is produced. But, as it follows this path an insect is either showered with pollen, or brushed by stigmas which pick up the pollen deposited on its body by another flower (Fig. 2).

Fertilization

Fertilization occurs after pollination. The nucleus of a pollen grain has to fuse with the nucleus of a female sex cell. This happens as follows. The pollen grain produces a long tube (Fig. 3B). This tube grows down into the ovule where its tip bursts open. A nucleus travels down this tube into the ovule where it fuses with the nucleus of the female sex cell.

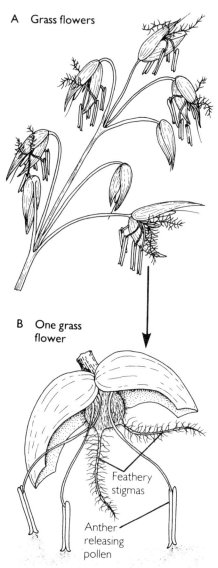

A   Grass flowers

B   One grass flower

Feathery stigmas

Anther releasing pollen

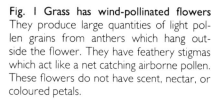

Fig. 1 Grass has wind-pollinated flowers They produce large quantities of light pollen grains from anthers which hang outside the flower. They have feathery stigmas which act like a net catching airborne pollen. These flowers do not have scent, nectar, or coloured petals.

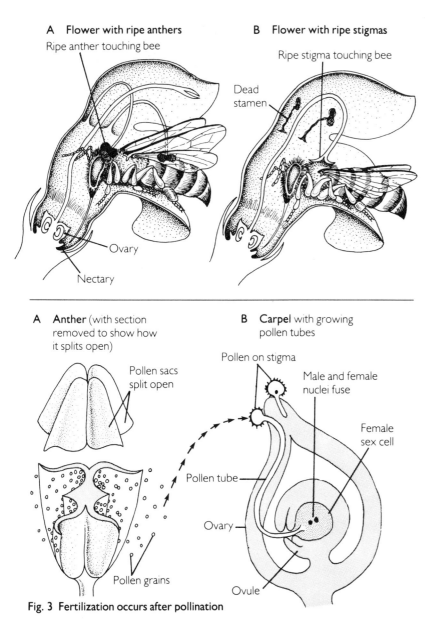

A Flower with ripe anthers

Ripe anther touching bee

Ovary

Nectary

B Flower with ripe stigmas

Ripe stigma touching bee

Dead stamen

Fig. 2 White dead nettle has insect-pollinated flowers Flowers of this type attract insects by scented petals which are often brightly coloured, and reward insects with nectar and pollen. Anthers and stigmas often ripen at different times. This prevents self-pollination.

A Anther (with section removed to show how it splits open)

Pollen sacs split open

Pollen grains

B Carpel with growing pollen tubes

Pollen on stigma

Male and female nuclei fuse

Female sex cell

Pollen tube

Ovary

Ovule

Fig. 3 Fertilization occurs after pollination

An insect-pollinated flower (daisy)

Exercises

1 What is the difference between self-pollination and cross-pollination?

2 How do the following increase the chances of pollination in grass flowers:
   a) production of pollen in large quantities?
   b) small, light-weight pollen grains?
   c) spreading stigmas?
   d) flowers on long stalks well above the leaves?
   e) anthers on long stalks?

3 Describe the ways in which insect-pollinated flowers attract insects.

4 Insect-pollinated flowers produce fewer and larger pollen grains than wind-pollinated flowers. Explain why this is so.

5 Describe how, after pollination, the nucleus of a pollen grain reaches the nucleus of the female sex cell.

A wind-pollinated flower (willow catkins)

8.4 Seeds, dispersal, and germination

Seeds appear to be hard, dry, and lifeless. But a spark of life is preserved within them for months, or even years, until conditions are right for growth. There are unconfirmed reports that corn seeds grew successfully after being stored in an Egyptian tomb for thousands of years.

Parts of a seed

A seed is made up of a small undeveloped plant, the **embryo**, enclosed in a protective skin along with a supply of food. When it grows, or **germinates**, the embryo plant uses up the stored food to reach the stage at which it can make its own food by photosynthesis.

Bean and pea seeds have food stored in swollen seed leaves of the embryo itself (Fig. 1B). In wheat, maize, and other cereals, some food is stored in the embryo and some is in a compartment next to the embryo (Fig. 1C). This food is called **endosperm**. It is released during milling and is commonly known as flour.

Dispersal

It is better that seeds begin to germinate some distance from the parent plant. Otherwise overcrowding will occur, and there will be little chance of the plants spreading into new territory. Seeds are dispersed by wind and by animals (Figs. 2 and 3).

Germination

The conditions in which successful germination occurs are illustrated in Figure 4.

Exercises

1　Name the parts of a seed. What are the main differences between a bean seed and a barley grain?

2　Why is it important that seeds grow some distance away from their parent plant?

3　What is endosperm? Why is endosperm useful for humans as well as seeds?

4　Explain what germination is.

5　What are the features of A, B, and C, of Figure 2 which allow them to be carried away by wind?

6　What part do animals play in dispersing each of the fruits and seeds illustrated in Figure 3?

7　The experiment in Figure 4 shows that bean seeds require three things before they will grow. What are these three things?

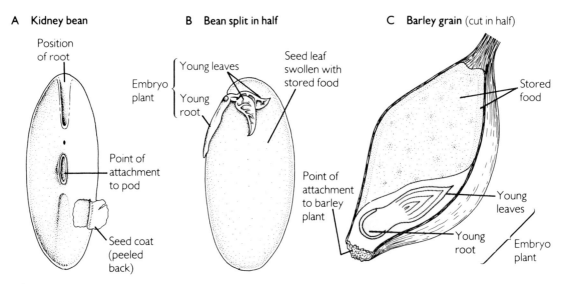

Fig. 1 Parts of a seed

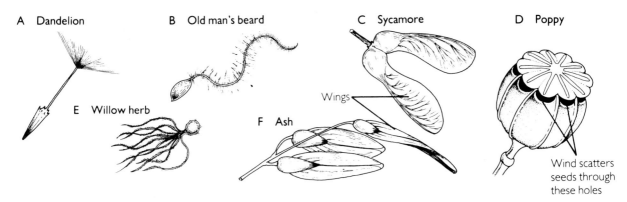

A Dandelion   B Old man's beard   C Sycamore   D Poppy

E Willow herb

F Ash

Wings

Wind scatters
seeds through
these holes

Fig. 2 Fruits and seeds dispersed by wind (exercise 3)

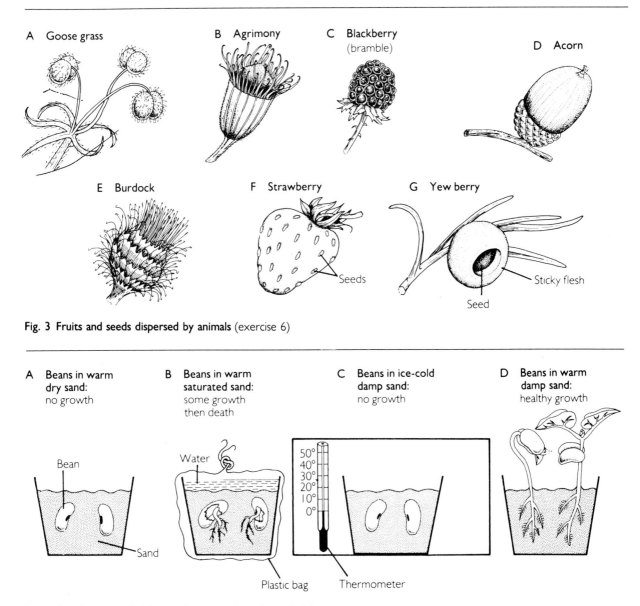

A Goose grass   B Agrimony   C Blackberry
(bramble)   D Acorn

E Burdock   F Strawberry   G Yew berry

Seeds

Sticky flesh

Seed

Fig. 3 Fruits and seeds dispersed by animals (exercise 6)

A **Beans in warm
dry sand:**
no growth

B **Beans in warm
saturated sand:**
some growth
then death

C **Beans in ice-cold
damp sand:**
no growth

D **Beans in warm
damp sand:**
healthy growth

Bean

Water

50°
40°
30°
20°
10°
0°

Sand

Plastic bag

Thermometer

Fig. 4 Conditions needed for seeds to germinate (exercise 7)

8.5   Insect reproduction

More than 80 per cent of all animals are insects, and the number alive at any one time cannot be counted or even estimated. Insects reproduce so quickly that if countless millions were not eaten each day by other animals they would soon strip the world bare of plants and many other forms of life.

Butterfly pupae

Eggs, nymphs, and adults

Dragonflies, grasshoppers, and bugs are examples of insects whose eggs develop into **nymphs**. A nymph is very similar to the adult it will become except that it is smaller, and lacks the wings and reproductive organs of an adult (Fig. 1B). Nymphs spend their time eating, growing, and moulting. With each moult they get bigger and their wings and their reproductive system are more fully developed.

Eggs, nymphs, and virgin births

Aphids (also called greenfly, or plant lice) reproduce all summer without a male to be seen. During this season of abundant food only females are to be found. They do not have to waste time finding a mate because their eggs develop without being fertilized. They do not have to lay their eggs because the eggs develop inside the female into nymphs which take little more than a week to become fully grown. Even this growth period is not wasted because, while she is still inside her mother, a female has eggs developing inside her own body! A female aphid can produce up to twenty-five nymphs a day and about one hundred in her lifetime. Male aphids are produced in the autumn. These mate with females, which lay eggs that survive the winter (Fig. 2).

Development without fertilization is called **parthenogenesis**, or virgin birth. In aphids it is important as a means of rapidly building up huge numbers of insects while food is still plentiful.

Eggs, larvae, pupae, and adults

Most types of insects lay eggs which hatch into **larvae**. A larva does not look at all like the adult it will become. A caterpillar, for example, does not look like the butterfly it will develop into (Fig. 3B). A larva eats and moults until it is full size, then it stops feeding and changes into a **pupa**, or **chrysalis** (Fig. 3C). Inside the pupa the larva is completely rebuilt to form the adult insect.

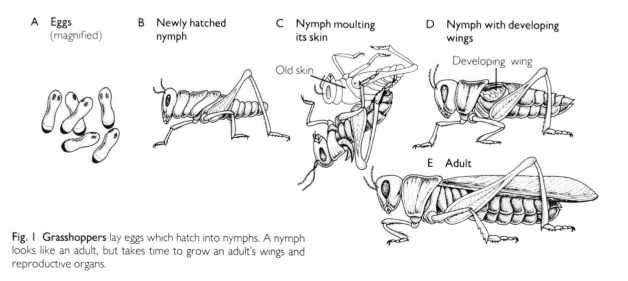

A **Eggs** (magnified) B **Newly hatched nymph** C **Nymph moulting its skin**   Old skin D **Nymph with developing wings**   Developing wing

E **Adult**

Fig. 1 Grasshoppers lay eggs which hatch into nymphs. A nymph looks like an adult, but takes time to grow an adult's wings and reproductive organs.

Exercises

Match the following words with the statements below:
nymph, moulting, larva, pupa, chrysalis, parthenogenesis.

a) Shedding the skin when an insect has grown too big for it.

b) A newly hatched insect which is similar to its parents.

c) Virgin birth.

d) Another name for a pupa.

e) A newly hatched insect which is completely different from its parents.

f) The stage during which a larva becomes an adult insect.

g) A caterpillar is an example.

h) Aphid females can reproduce this way.

Swallowtail butterfly emerging from its chrysalis

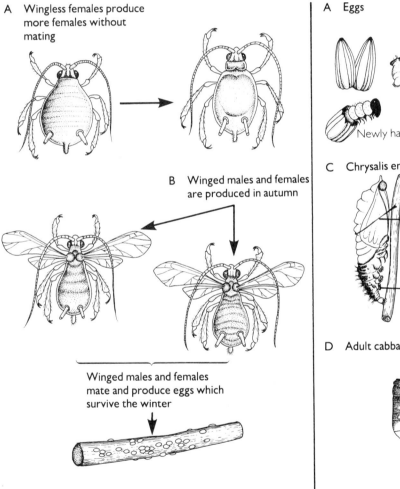

A Wingless females produce more females without mating

B Winged males and females are produced in autumn

Winged males and females mate and produce eggs which survive the winter

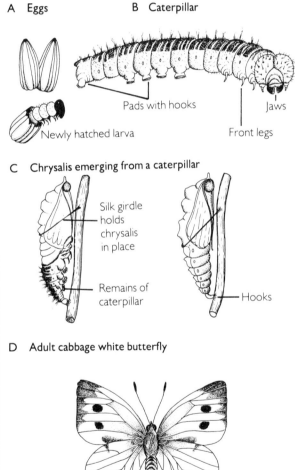

A Eggs   B Caterpillar

Pads with hooks   Jaws

Newly hatched larva   Front legs

C Chrysalis emerging from a caterpillar

Silk girdle holds chrysalis in place

Remains of caterpillar

Hooks

D Adult cabbage white butterfly

Fig. 2 Aphid (greenfly) eggs hatch into wingless female nymphs (**A**). Eggs inside the females develop without being fertilized. These females give birth to more females which repeat the process all summer. In autumn winged males and females are born (**B**). These mate and the females lay eggs which can survive the winter.

Fig. 3 A butterfly egg hatches into a larva (caterpillar). This looks nothing like the adult insect. Caterpillars feed for a time and then change into a chrysalis (pupa). The adult butterfly develops inside the chrysalis.

8.6   Human reproductive organs

How are eggs and sperms brought together in the Animal Kingdom? Wind and insects cannot help out with animal fertilization!

Types of fertilization

Figures 1A and B illustrate two examples of fertilization which occur outside a female animal's body. This type of fertilization is called **external fertilization**. These methods are common among fish and other water creatures. Large numbers of eggs and sperms must be produced because fertilization is chancy and many eggs are eaten by other animals.

Figure 1C illustrates **internal fertilization**. Some fish and all reptiles, birds, and mammals employ this method. It gives the best chance of fertilization because eggs and sperms are brought together in a confined space which gives protection.

Reptile and bird eggs are protected after they are laid by a shell. The eggs of humans and other mammals are not only fertilized inside the female, they develop there. This is called **internal development**. It means the young are protected, and kept warm and fed inside their mother until they are fully formed. After birth they are protected by their parents until they are old enough to look after themselves.

Male reproductive organs

Sperms are passed from a man to a woman by an organ called a **penis** (Fig. 2). When a man is sexually stimulated spaces inside his penis fill with blood making it firm and erect. During sexual intercourse he moves his penis back and forth inside a woman's **vagina** (Fig. 3). This eventually causes muscular contractions in the man's sperm tubes. These contractions propel a liquid called **semen** into the female. Semen contains millions of sperms together with chemicals which nourish them and cause them to begin swimming.

Female reproductive organs

Sexual intercourse usually causes muscular contractions in the woman's reproductive organs which suck sperms from her vagina through her **cervical canal** into her **uterus** (Fig. 3). From here sperms must swim up the **fallopian tubes** to where fertilization occurs.

Fertilization can only occur if a fallopian tube contains an egg cell (ovum) which has recently been released from the woman's ovaries. Fertilization is described in Unit 8.7.

A fertilized ovum moves down a fallopian tube to the uterus where it develops into a baby. Development and birth are described in Unit 8.8.

Key: Male
         Female

A   Random external fertilization
Eggs and sperms are released at random into water. Large quantities of each are produced since many sperms will fail to find an egg.

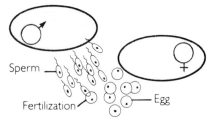

Sperm

Fertilization ———— Egg

B   External fertilization in a nest
Eggs and sperms are deposited into a specially prepared hollow or other nest. This greatly increases the chances of fertilization.

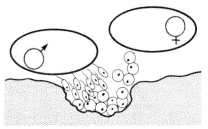

C   Internal fertilization   Sperms are passed into the female's body where fertilization occurs. Few eggs are produced since eggs and sperms are in a confined space.

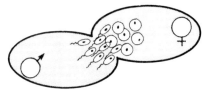

Fig. 1 Types of fertilization

Exercises

1. Match the following words with the sentences below: testis, ovary, uterus, penis, semen.
   a) A baby develops here.
   b) Produces sperms.
   c) Passes sperms into the female during sexual intercourse.
   d) Produces ova.
   e) A liquid containing sperms.

2. a) What are the disadvantages of external fertilization?
   b) What are the advantages of internal fertilization, and of internal development?
   c) List some animals which employ: external fertilization, internal fertilization, internal development.

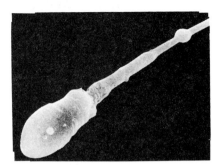

Highly magnified human sperm

A Front View

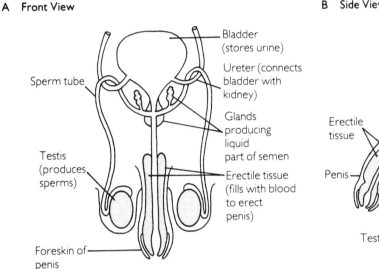

B Side View

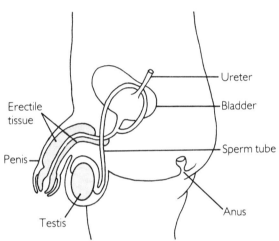

Fig. 2 Human male reproductive organs

A Front View

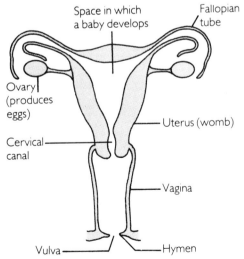

B Side View

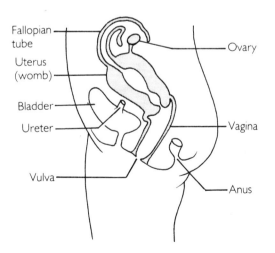

Fig. 3 Human female reproductive organs

8.7   How a new life begins

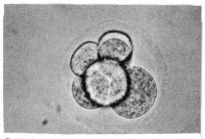

Only 5 cm³ of semen enter a woman's reproductive system during sexual intercourse but this liquid contains up to 300 million sperms. Yet only one sperm is needed to fertilize an ovum and start the development of a new human life.

Cells dividing in an early human embryo

Female periods and the menstrual cycle

Fertilization can only occur if an ovum is in a woman's fallopian tubes when sexual intercourse takes place. An ovum is released from an ovary once a month as part of a sequence of events called the *menstrual cycle*. This sequence of events takes about 28 days and is repeated continuously from the time a girl becomes sexually mature (between eight and fifteen years of age) until she is aged about forty-five to fifty. This cycle stops during pregnancy and starts again soon after the baby is born.

Events during one menstrual cycle

1   During the first five days of a cycle the uterus loses its lining. The lining breaks down and passes out of the vagina with a quantity of blood. This is *menstruation*, and is commonly known as *a period* (Fig. 1A).

2   Between about the thirteenth and fifteenth days one ovary releases an ovum. This is called *ovulation* (Figs. 1B and 2A). The ovum travels slowly along a fallopian tube towards the uterus.

3   Between ovulation and the end of the cycle on the twenty-eighth day the uterus grows a new lining. This lining consists of glands and blood vessels and is needed to nourish the ovum in case it is fertilized and develops into a baby.

4   If the ovum is not fertilized menstruation begins again on or about the twenty-eighth day of the cycle (Fig. 1).

Fertilization and the start of development

The first sperm to reach an ovum burrows into it so that the sperm's nucleus can fuse with the ovum's nucleus (Fig. 2B and C). At this moment a skin forms around the ovum preventing the entry of other sperms.

The fertilized ovum begins to divide and by the time it reaches the uterus it is a ball of cells called an *embryo* (Fig. 2E and F).

This ball becomes firmly embedded in the uterus wall (Fig. 2G and H). Here it receives food and oxygen from the mother's blood supply and grows into a baby.

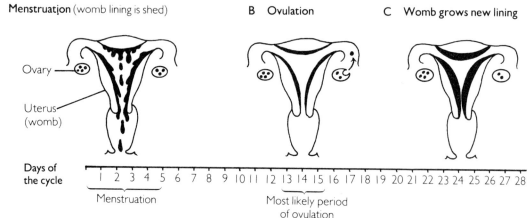

A   Menstruation (womb lining is shed)   B   Ovulation   C   Womb grows new lining

Ovary

Uterus (womb)

Days of the cycle   1 2 3 4 5 6 7 8 9 10 11 12 13 14 15 16 17 18 19 20 21 22 23 24 25 26 27 28

Menstruation   Most likely period of ovulation

Fig. 1 The menstrual cycle is a sequence of events repeated monthly in the female reproductive system.

Exercises

1 a) How many sperms are needed to fertilize an ovum?
   b) Describe what happens when a sperm and an ovum fuse together.

2 What changes take place in a fertilized ovum as it travels to the uterus, and what happens when it gets there?

3 What is menstruation and what is its common name?

4 a) List the events which occur during one menstrual cycle.
   b) What is the purpose of these events?

5 What is implantation (Fig. 2)?

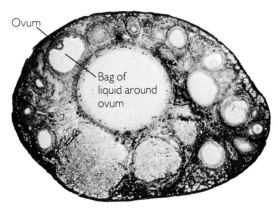

Highly magnified human ovary (cross section)

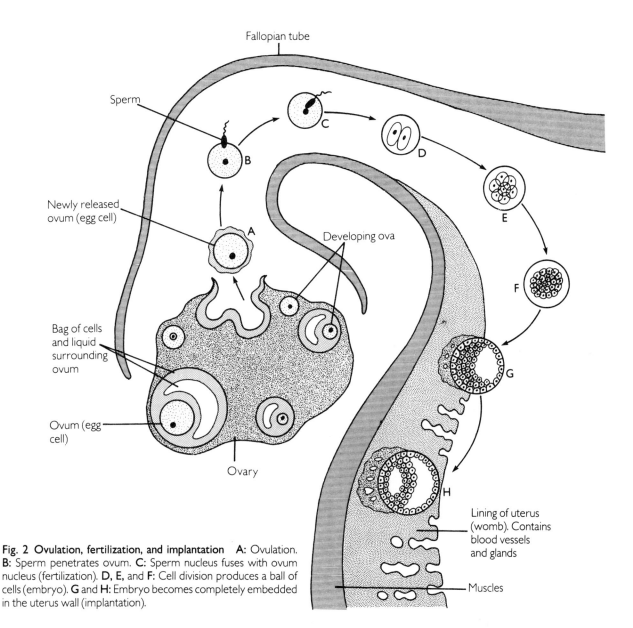

Fig. 2 Ovulation, fertilization, and implantation A: Ovulation. B: Sperm penetrates ovum. C: Sperm nucleus fuses with ovum nucleus (fertilization). D, E, and F: Cell division produces a ball of cells (embryo). G and H: Embryo becomes completely embedded in the uterus wall (implantation).

8.8   Development and birth of a baby

In nine months a single cell, the fertilized ovum, one-tenth of a millimetre across, becomes a fully formed baby made up of billions of cells organized into scores of different tissues and organs.

The placenta

A placenta is a disc-shaped organ attached to the wall of the womb. It is connected with the embryo by a tube called the **umbilical cord** (Figs. 1 and 2).

The developing baby's heart pumps blood through the umbilical cord into the placenta. Here the baby's blood absorbs food and oxygen *from* its mother's blood, and releases carbon dioxide and other wastes *into* its mother's blood.

In other words the placenta does jobs which will eventually be carried out by the baby's digestive system, lungs, and kidneys.

The amnion

A baby develops inside a bag of liquid called an **amnion** (Fig. 1). Liquid in the amnion cushions the baby against jolts and shocks as its mother moves about.

Development and birth

Photos of a developing baby are shown in Figure 2. Figure 3 illustrates the birth of a baby. After the baby has been born further contractions of the womb push the placenta and umbilical cord out of the mother's body. These are called the **after-birth**.

Exercises

1   What happens to a baby's blood as it passes through a placenta? Which organs will carry out the placenta's functions after a baby is born?

2   What does an umbilical cord contain (Fig. 1)?

3   What is an amnion, and what is the function of the liquid inside an amnion?

4   Describe the main differences between a 40-day embryo and a foetus (Fig. 2).

5   What causes labour pains (Fig. 3)?

6   What is the after-birth?

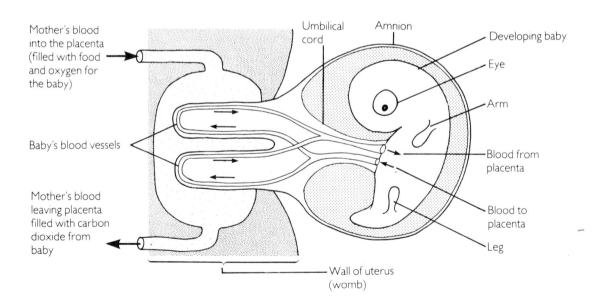

Fig. 1 **A developing baby** with placenta and amnion.

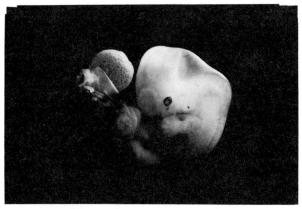

This embryo is 40 days old. It is about 1.5 cm long. The umbilical cord is forming and five fingers are faintly visible on the hand. The nose and cheeks are beginning to form and the eye is developing.

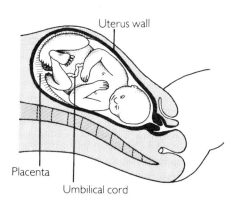

A A few days before birth a baby turns in the womb until its head points downwards. Birth begins when the womb starts contracting and relaxing rhythmically. These contractions cause labour pains.

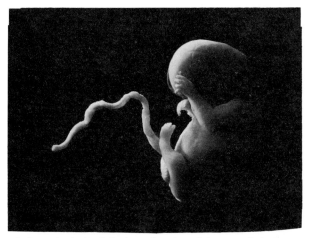

At 9 weeks the body is about 4 cm long. It is now called a foetus. The body is almost completely formed.

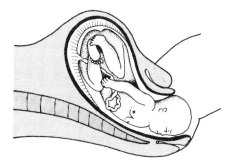

B Contractions of the womb push the baby out of the mother's body

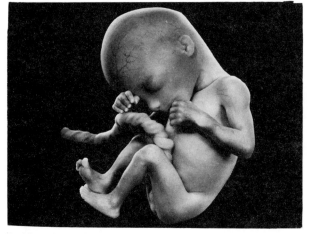

At 16 weeks the foetus is about 16 cm long. It will continue to grow inside the womb until birth.

Fig. 2 Development of a baby

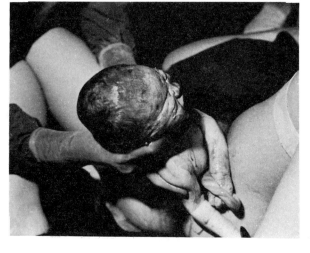

Fig. 3 Birth of a baby

8.9 Twins, triplets, and inherited characteristics

Usually a woman's ovaries release one ovum a month, but sometimes more than one is released. Furthermore, after an ovum has been fertilized it can sometimes divide into completely separate cells. These are the ways in which twins, triplets, quadruplets, etc. are produced.

Identical and non-identical twins

Sometimes a fertilized ovum produces more than one baby. This can happen if, after it has been fertilized, the ovum divides not into one ball of cells, but into two or more (Fig. 1A). If this happens the babies so produced will be identical to one another, because they are all from the same ovum and sperm.

Sometimes, however, a woman's ovaries release two or more ova at the same time (Fig. 1B). If all these ova are fertilized they will develop into babies which are different from one another. This happens because each baby has developed from a different ovum, and each ovum was fertilized by a different sperm.

Inherited and acquired characteristics

Identical twins were separated soon after birth and brought up in different countries, but met each other again twenty years later. Imagine them standing side by side. In what ways would they still be identical, and how could they be different from each other?

Features such as the shape of the face, ears, nose, and mouth, and colour of the skin, hair, and eyes should still be identical although some of these features could have been altered either accidentally or deliberately. But their language, skills, etc., will probably be different.

The twins inherited all their identical features such as eye and hair colour from their parents. Consequently these are called *inherited characteristics*. Features like their language and skills are called *acquired characteristics*, because the twins acquired them during their lives. These features depend upon where and how a person lives.

A **Identical twins** are formed when a fertilized egg splits into two cells each of which grow into an embryo

B **Non-identical twins** are produced when two egg cells are released at the same time from each ovary and both are fertilized

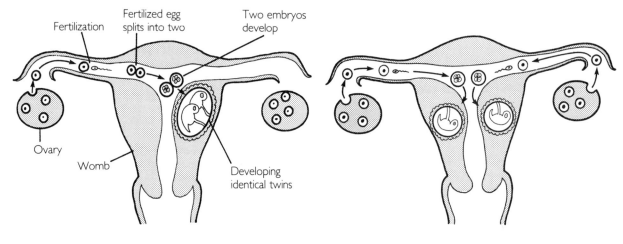

Fig. 1 How twins are produced

Exercises

1 Study Figure 2 and make a list of the *visible* inherited and acquired characteristics of the twins. Make a list of other inherited and acquired characteristics which they may have, but which are not visible.

2 A man has black hair and brown eyes, he speaks Spanish, and has a finger missing. Which of these character- istics could his children inherit from him?

3 Describe how identical and non-identical twins are formed.

4 Anne, Christine, and Janet are triplets. Christine and Janet are identical. Describe how these triplets could have been formed.

Fig. 1 Identical twins which were separated soon after birth, brought up in different countries, but reunited twenty years later. (See exercise 1.)

8.10   Genetics and genes

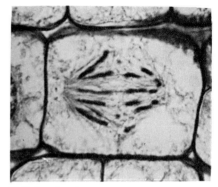

Cell showing chromosomes

Mother has brown eyes and father's eyes are blue. Will their children have one brown eye and one blue eye, or will their eyes be a muddy coloured mixture of brown and blue?

Genetics—the scientific study of inheritance

Children cannot inherit *all* the characteristics of *both* their parents. They inherit some characteristics from each parent. In the example above it is likely that some children will have brown eyes and others blue eyes. The science of genetics tells us why.

Cells, chromosomes, and genes

The nucleus of a living cell contains long, thin strands called *chromosomes* (Fig. 1A). Each chromosome is made up of smaller parts called *genes*. Genes are arranged along the chromosome as shown in Figure 1B. Genes control the development of inherited characteristics. For example, different genes control the development of eye colour, hair colour, the shape of the nose, mouth, ears, etc.

A sperm contains chromosomes with a set of genes from the male parent. An ovum contains chromosomes with a set of genes from the female parent. At fertilization these two sets of genes are brought together. They then control development of the fertilized ovum into a baby.

Dominant and recessive genes

Some genes are said to be *dominant* because they have a stronger control over development than other, *recessive* genes. If, for example, the gene controlling development of brown eyes is brought together in a fertilized ovum with a blue-eyed gene, then the fertilized ovum will become a brown-eyed child. Brown eye colour is called a *dominant characteristic* and blue eye colour is a *recessive characteristic*. Figures 2 and 3 illustrate other dominant and recessive characteristics.

Exercises

1   Whereabouts in a cell are chromosomes found?

2   Where are genes found?

3   What is the function of genes?

4   Pollen from a pea plant which grows green pods was used to pollinate flowers of a pea plant with yellow pods. Seeds from the pollinated plant *all* grew into plants with green pods.
   a) What technical words should be used to describe the two pod colours?
   b) Both types of plant used in this experiment were pure bred. What does this mean (Fig. 3A)?
   c) Try to construct a diagram like Figure 3 to illustrate how the green pods were produced from green and yellow podded parents.

A   **Diagram of a cell**   (The nucleus is shown larger than life)

Chromosomes (there are 46 in human cells)

Nucleus

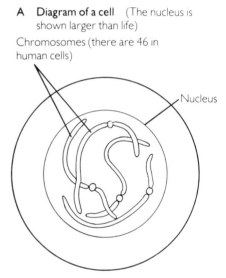

B   Diagram of a chromosome

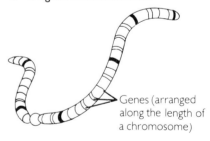

Genes (arranged along the length of a chromosome)

Fig. 1 **The nucleus** of a cell contains chromosomes, and these are made up of genes. Genes control the development of inherited characteristics.

Dominant characteristics	**Recessive characteristics**

Ear lobes present

Ear lobes absent

Straight nose

Upturned nose

Dark hair and dark eyes

Light hair and light eyes

Tallness and coloured flowers in pea plants

Dwarfness and white flowers in pea plants

A   Parents

Male – a pure-bred, dark-coated mouse (*only* dark-coated mice among its ancestors)

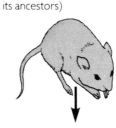

Female – a pure-bred, light-coated mouse (*only* light-coated mice among its ancestors)

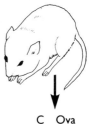

B   Sperms

All sperms will have the gene which controls dark coat

C   Ova

All ova will have the gene which controls light coat

Chromosome

Gene for dark coat

Gene for light coat

D   Fertilization (dark and light genes brought together)

Fertilized ova contain the dark and light genes

E   Young

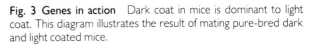

All the young have dark coats

Fig. 2 Dominant and recessive characteristics   If one parent has a dominant characteristic and the other has the opposite recessive characteristic, the dominant one will appear in at least half of their young.

Fig. 3 Genes in action   Dark coat in mice is dominant to light coat. This diagram illustrates the result of mating pure-bred dark and light coated mice.

8.11   New medical techniques

Louise Brown, the world's first test tube baby, was born on 25 July 1978. What benefits and problems will new medical techniques bring the human race?

What is a test tube baby?

Some women who are unable to have a baby now have the chance to produce one using the so-called test tube method. The baby does not develop in a test tube. One or more fully developed egg cells are sucked from a woman's ovaries through a special syringe inserted through her abdomen. The eggs are placed in a dish or tube with sperms from her husband, and kept warm for about 60 hours. During this time sperms fertilize the eggs and the eggs divide forming embryos. One embryo is then inserted into the woman's womb where it develops normally (Fig. 1).

Storage of embryos   One day it will be possible to store embryos indefinitely, probably by freezing them. A couple could then have several eggs fertilized at one time and store the resulting embryos. Whenever they wanted a child an embryo could be placed in the woman's womb. But what would be done with unwanted embryos? Would it be murder to throw them away when their parents die? Does an embryo have the rights of a human being?

Baby factories   Scientists are already trying to progress from test tube fertilization to full development of babies in an artificial womb. An artificial womb would have to supply oxygen and food to the baby and remove its waste. Thus, it would do the jobs normally done by the mother's lungs, digestive system, liver, kidneys, and bloodstream.

Parents would be able to watch their baby grow into a boy or girl, and birth would take place without pain or effort on the mother's part. But would she truly feel she was its mother?

Carbon copy people

Any number of identical plants can be produced from a single parent by taking cuttings. Eventually scientists will be able to produce countless identical animals, even people, using cells taken from single individuals. Figure 2 explains how this has been done using frogs. These exact 'carbon copies' or *clones* will have all the hereditary features of their parent. In other words they will not only be identical twins of each other but of their parent too!

Animal clones   It would obviously be profitable to make exact copies of prize bulls, cows, sheep, race-winning horses, etc. But is there any point in copying people?

Human clones   Identical (one-egg) twins are often more aware of each other's needs and problems than non-identical brothers and sisters. If clones have this close relationship they could make world-beating sports

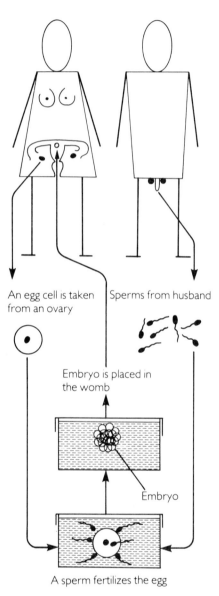

An egg cell is taken from an ovary

Sperms from husband

Embryo is placed in the womb

Embryo

A sperm fertilizes the egg

Fig. 1 Test tube babies are produced by taking an egg from a woman and placing it in a container with sperms from her husband. If the egg is fertilized the resulting embryo is placed in the woman's womb to develop normally.

teams, spacecraft crews, exploration teams, etc. Remember, however, that although clones will have identical bodies, their personalities and intelligence could differ. Consequently, an exact copy of a scientific genius, artist, or military leader, might not share its parent's abilities.

Tinkering with heredity

Figure 3 describes how scientists can change the very stuff of life itself—*genes*, which control development of hereditary characteristics.

Louise Brown, the world's first test tube baby, with her parents.

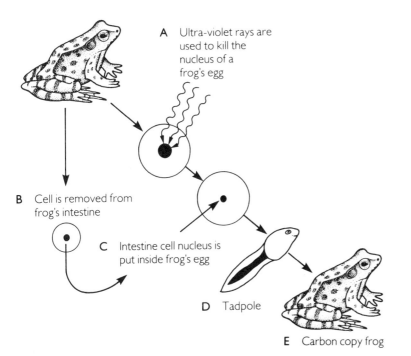

A Ultra-violet rays are used to kill the nucleus of a frog's egg

B Cell is removed from frog's intestine

C Intestine cell nucleus is put inside frog's egg

D Tadpole

E Carbon copy frog

Fig. 2 Carbon copy frogs It is possible to make a frog with only one parent. **A:** First, an egg cell nucleus is destroyed. **B** and **C:** This nucleus is then replaced with one from a frog's body cell (e.g. from the intestine). The egg cell now has exactly the same hereditary material as the parent frog and so develops into an exact copy of that frog (**D** and **E**). The frog's parent is itself! This technique could be used to copy prize-winning cattle, sheep, horses, etc., and even people.

Fig. 3 Tinkering with heredity This is a bacterium surrounded by chains of genes (its own hereditary material). Scientists can now add new genes to bacteria to make them produce antibiotic drugs and other useful chemicals; to make them gobble up spilled oil; and to make them able to live inside plant roots and produce fertilizer free of charge. One day 'bad' genes could be cut out of human cells to cure hereditary diseases like haemophilia.

8.12 Regeneration: new parts from old

An earthworm can replace a lost end, lobsters can replace lost limbs, and plant cuttings can grow new roots, leaves, and a stem. Our bodies can heal cuts and broken bones, and recent experiments suggest that one day we could grow replacement arms, legs, or even internal organs.

Regeneration in plants and animals

The power to regenerate new parts depends upon the ability of undamaged cells in the injured area to produce the kinds of cells which have been lost.

Regeneration in plants   Young plants and the growing parts of old plants contain cells which have not yet become specialized for a particular job. If you cut up a plant, these unspecialized cells can grow the missing parts. Pieces of root or stem, buds, or even a single leaf can grow into complete plants (Fig. 1).

Regeneration in animals   The less specialized an animal the more able it is to replace missing parts. In other words, regeneration is commonest in animals with a simple structure. Sponges, and the relatives of *Hydra* for example, are so unspecialized that they can be mashed up into separate cells and the cells will clump together again to form a new animal.

Flatworms called planarians can regenerate from tiny pieces (Fig. 2B and C), and if a predator bites all the arms off a starfish the central disc not only continues to feed but grows a new set of arms (Fig. 2A). Regeneration is rare among vertebrates. This may be because their cells become so specialized for a particular job that they lose the ability to produce more new cells. Young newts, frogs, and salamanders, however, can sometimes grow replacement limbs (Fig. 3), and a lizard can regenerate a lost tail but the new one is rarely full size.

Regeneration in humans

If a newt can grow a new leg why can't you? The answer may be that you could if you imitated newts. Recent research has shown that newts have a weak, natural, electrical current flowing through their skins which becomes fifty times stronger around the stump of a cut limb.

Researchers have had success using mild electric currents to speed up the healing of broken bones, and of cut nerves and muscles in human limbs. The current seems to stimulate the body's natural repair mechanism to such an extent that, if the ideal conditions could be found, researchers believe that whole replacement limbs, eyes, kidneys, and other organs could be persuaded to grow.

A   Stem cuttings Geraniums and coleus can be grown from pieces of stem. Put them in compost in a pot and cover with a plastic bag

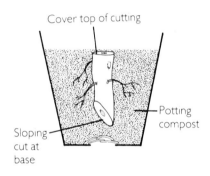

Remove leaves from lower third of cutting

Cut stem with a sharp knife below a leaf joint

B   Root cuttings Hollyhocks and phlox can be grown from pieces of root

Cover top of cutting

Potting compost

Sloping cut at base

C   Leaf cuttings *Begonia rex* can be grown from a leaf. **1.** Turn leaf upside down and cut across veins. **2.** Put leaf right side up onto compost and peg it in place. **3.** Small plants should grow from cuts.

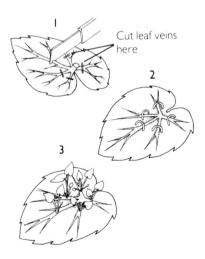

Cut leaf veins here

Fig. 1 Regeneration in plants

Exercises

1 What is regeneration?

2 How do gardeners use regeneration in plants?

3 Cut a side shoot off a geranium plant. The shoot should be about 10 cm long and the cut made just below a leaf (Fig. 1A). Cut off the lowest leaf. Put the cutting in a glass of water (changed once a week) and watch for regeneration of roots.

4 a) What is meant by an unspecialized animal? Name a few examples.

b) Why is regeneration commoner in unspecialized animals than in specialized ones?

5 Describe some examples of regeneration in the human body (Fig. 4).

A If a predator bites the arms off a starfish the central disc can grow new ones

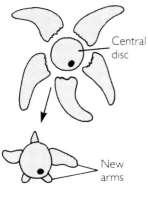

Central disc

New arms

B A whole new flatworm can regenerate from a tiny portion provided that portion contains gut and skin cells

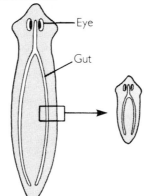

Eye

Gut

C The head of a flatworm cut down the middle can regenerate the missing parts

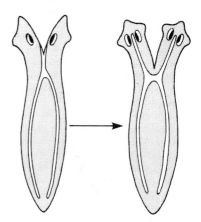

Fig. 2 Flatworms and starfish can regenerate from certain tiny pieces.

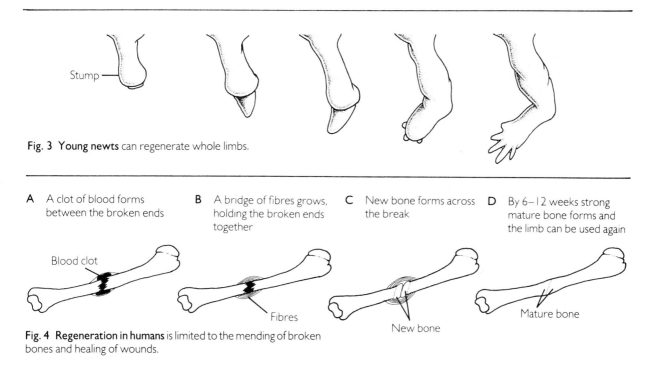

Stump

Fig. 3 Young newts can regenerate whole limbs.

A A clot of blood forms between the broken ends

Blood clot

B A bridge of fibres grows, holding the broken ends together

Fibres

C New bone forms across the break

New bone

D By 6–12 weeks strong mature bone forms and the limb can be used again

Mature bone

Fig. 4 Regeneration in humans is limited to the mending of broken bones and healing of wounds.

Topic 8 exercises

1 Write the scientific name for each of the following:
   a) Reproduction in which there is only one parent.
   b) Fusion of a sperm with an ovum.
   c) The cell produced by the fusion of a sperm and an ovum.
   d) The organs which produce sperms, pollen grains, and ova.
   e) Caterpillar and chrysalis.
   f) A human baby eleven weeks after fertilization of the ovum from which it developed.
   g) A woman's period.

2 a) List the differences between a sperm and a pollen grain.
   b) What are the differences between the ways pollen grains and sperms travel to and fertilize their female sex cells?

3 Compare the eggs of frog and bird illustrated in Figure 1.
   a) Why must a frog's eggs be laid in water?
   b) What feature of a bird's egg allows it to survive on land?
   c) Why must birds use internal fertilization?

4 A hen's egg is thousands of times larger than a human egg.
   a) What substances are present in large quantities in a hen's egg which account for this size difference?
   b) Why must hen's eggs have these substances in such large quantities?
   c) Why doesn't a human egg require a shell?
   d) How are human and hen's eggs kept warm while they develop?

5 a) What do the words 'larva' and 'nymph' mean?
   b) Sort the young animals in Figure 2 into larvae and nymphs.
   c) List the main differences between the young and adult of each animal illustrated in Figure 2.

Fig. 1 Frog and bird eggs   See exercise 3.

Frog

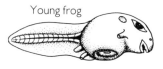
Young frog

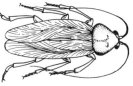
Cockroach

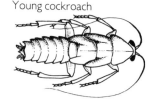
Young cockroach

Housefly

Young housefly

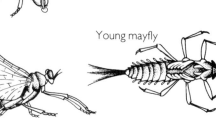
Mayfly

Young mayfly

Fig. 2 Nymphs, larvae, and adults   See exercise 5.

Health and hygiene

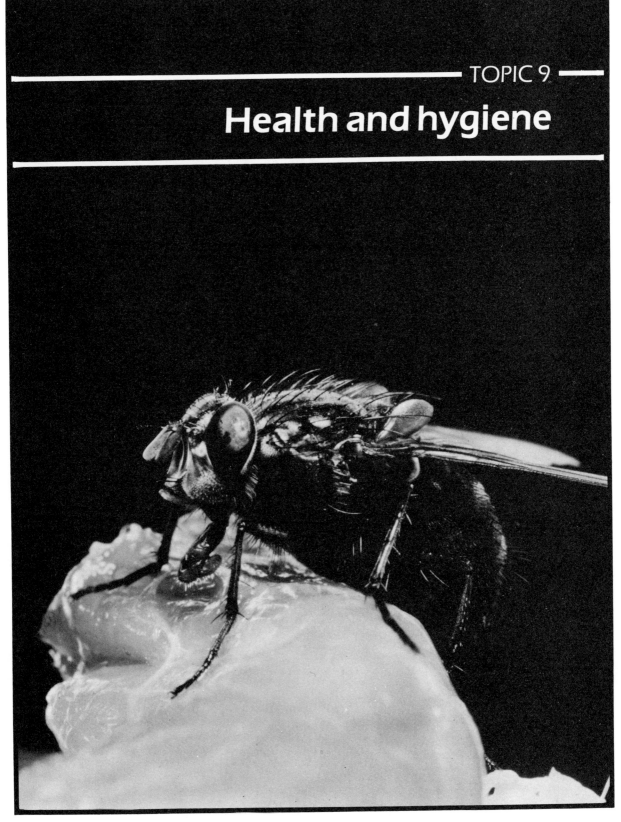

A blowfly getting its share of some meat

9.1 Microbes, germs, and disease

Some are only one-fifty-thousandth of a millimetre long, some eat crude oil and reproduce every 20 minutes, others can live for 600 million years and can survive a dose of radiation 10 000 times greater than that which would kill humans. What are they?

Microbes

The term microbe includes viruses, bacteria, protozoa, and certain fungi. Microbes which cause disease are commonly known as germs.

Viruses

Viruses are extremely small. It would take 6 million of the smallest viruses to make a row 1 mm long.

All viruses are parasites and only show signs of life when inside the living cells of a host. Here they plunder the cell for materials to make new viruses which are released and repeat the process in other cells. Outside their host cells viruses become dead chemicals. They can be dried into a powder and stored in a bottle for years and yet come to life again when placed on living tissue (Fig. 1).

Bacteria

Bacteria vary in size from 0.0005 mm to 0.005 mm. Bacteria live almost everywhere: in the sea where some can digest spilled crude oil, in boiling volcanic springs, in rotting flesh, and floating in the upper atmosphere.

In good conditions bacteria reproduce asexually every 20 minutes by splitting in two. When food is scarce they form spores which are incredibly tough. Some spores can resist boiling, drying up, intense radioactivity, and strong disinfectants. Living bacterial spores have been found embedded in layers of salt 600 million years old.

Harmful bacteria Boils, food poisoning, whooping cough, leprosy, diphtheria, and venereal disease are examples of diseases caused by bacteria (Fig. 1).

Useful bacteria Many bacteria are very useful. They are used to ripen cheese, to make sewage harmless, to produce vinegar, alcohol, and acetone, to produce silage (cattle food), and to ret flax (soften the plant fibres used to make linen cloth).

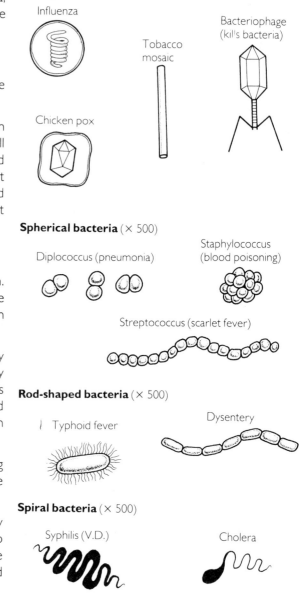

Viruses ($\times$ 10 000)

Influenza

Tobacco mosaic

Bacteriophage (kills bacteria)

Chicken pox

Spherical bacteria ($\times$ 500)

Diplococcus (pneumonia)

Staphylococcus (blood poisoning)

Streptococcus (scarlet fever)

Rod-shaped bacteria ($\times$ 500)

Typhoid fever

Dysentery

Spiral bacteria ($\times$ 500)

Syphilis (V.D.)

Cholera

Fig. 1 **Some examples of microbes** – viruses and bacteria

Protozoa

Protozoa are single-celled organisms (Fig. 2). Amoebic dysentery, malaria, and sleeping sickness are examples of diseases caused by protozoa.

Fungi

Very few fungi are parasites of humans. Those which are include ringworm which causes circular swellings on skin, and a disease called athlete's foot which attacks soft skin between the toes (Fig. 2). Fungi cause many plant diseases. Mildew is an example (see p. 63).

Exercises

1 a) What do the words 'microbe' and 'germ' mean?
   b) Name a few germs, and name a microbe which is not a germ.
2 a) How do bacteria reproduce asexually?
   b) In ideal conditions how many bacteria could a single bacterium produce in 24 hours?
3 What is the size in micrometres of: a white blood cell, a typhoid bacterium, and an influenza virus (Fig. 3)?
4 List some of the differences between viruses and bacteria.
5 In what ways are bacteria useful?
6 Which disease can you avoid by washing your feet regularly and carefully drying between your toes?

Protozoa

Amoeba which causes dysentery ($\times$ 150)

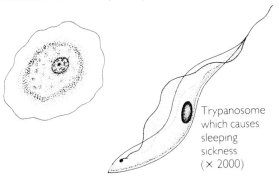

Trypanosome which causes sleeping sickness ($\times$ 2000)

Ringworm fungus ($\times$ 50)

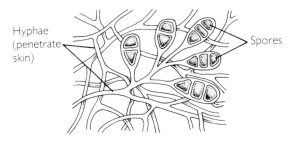

Hyphae (penetrate skin)

Spores

Fig. 2 **Some more examples of microbes** – protozoa and fungus

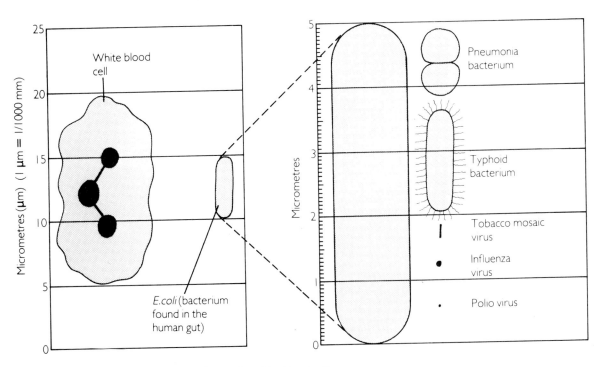

Fig. 3 Compare the sizes of these microbes with a white blood cell

9.2 Avoid dirt and germs

Sneezes from infected people can hurl germ-laden droplets of moisture a distance of more than four metres. So sneeze into disposable tissues, then burn them!

The spread of infection

The main sources of infection are: contact with infected people or animals, or with objects they have contaminated with germs, and consumption of contaminated food or drink (Fig. 1).

Contact You can catch ringworm and athlete's foot, chicken-pox, measles, boils, and septic wounds either by direct contact with infected people, or by touching objects they have touched like towels, combs, lavatory seats, and crockery (especially if cracked or chipped). Colds, influenza, pneumonia, diphtheria, whooping cough, and tuberculosis are spread by coughs and sneezes. Some germs, like venereal disease bacteria, quickly die outside the body, and so can *only* be spread by direct contact with an infected person.

Food and drink Food can be covered with germs by coughs and sneezes, by unwashed hands or hands with uncovered septic wounds, by flies, pets, and wild mice and rats. In countries with inefficient sewage disposal arrangements, drinking water can be contaminated with urine and faeces and should always be boiled before use.

These maggots have hatched from eggs which a fly laid in the meat

Personal cleanliness

Are you nice to know or are your clothes, hair, and feet so smelly that even dogs avoid you? Personal cleanliness is important in the fight against infection. Figure 2 illustrates the most important facts of personal cleanliness.

Houseflies live, breed, and feed on decaying matter and dung. Their legs, bodies, and gut become contaminated with germs which can include: typhus, cholera, dysentery, tuberculosis, and anthrax, together with certain tapeworm eggs. If allowed to settle on food they spread these germs and eggs over it as they walk and feed.

Fig. 1 The spread of infection

Terrapins spread germs which cause food poisoning.

Dogs can catch rabies from wild animals and spread it to humans. They can also spread certain tapeworm eggs.

Septic wounds and boils can be spread by direct contact, or by contact with towels, combs, cups, food, etc. which infected people have touched.

Coughs and sneezes spread colds, influenza, pneumonia, diphtheria, whooping cough, and tuberculosis.

Exercises

1   List some of the germs you could pick up from a dirty towel used by other people.

2   Why must you refuse to use a cracked or chipped cup given to you in a restaurant?

3   Why should people with coughs and sneezes stay at home until they are better?

4   When and why should you wash your hands?

5   a) Where do the germs come from which contaminate a housefly's body?
    b) How are these germs passed on to humans?

6   a) List some of the ways in which food and drink can be contaminated with germs.
    b) List some of the ways in which food and drink can be kept free of germs.

7   Why is it important to:
    a) wash off cosmetics with good quality soap each night?
    b) avoid excessive use of anti-perspirants in hot weather (see Unit 5.6)?
    c) wash your feet and change your socks daily?
    d) avoid using a comb which others have used before you?
    e) use only good quality sunglasses?

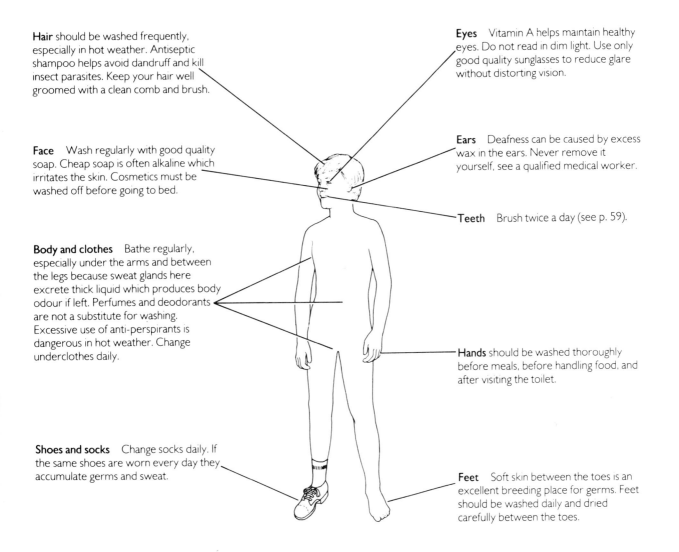

Hair should be washed frequently, especially in hot weather. Antiseptic shampoo helps avoid dandruff and kill insect parasites. Keep your hair well groomed with a clean comb and brush.

Face   Wash regularly with good quality soap. Cheap soap is often alkaline which irritates the skin. Cosmetics must be washed off before going to bed.

Body and clothes   Bathe regularly, especially under the arms and between the legs because sweat glands here excrete thick liquid which produces body odour if left. Perfumes and deodorants are not a substitute for washing. Excessive use of anti-perspirants is dangerous in hot weather. Change underclothes daily.

Shoes and socks   Change socks daily. If the same shoes are worn every day they accumulate germs and sweat.

Eyes   Vitamin A helps maintain healthy eyes. Do not read in dim light. Use only good quality sunglasses to reduce glare without distorting vision.

Ears   Deafness can be caused by excess wax in the ears. Never remove it yourself, see a qualified medical worker.

Teeth   Brush twice a day (see p. 59).

Hands should be washed thoroughly before meals, before handling food, and after visiting the toilet.

Feet   Soft skin between the toes is an excellent breeding place for germs. Feet should be washed daily and dried carefully between the toes.

Fig. 2 Personal cleanliness helps fight infection

9.3 Smoking kills

Most of the ill-health people suffer could be prevented if they ate balanced meals, took regular exercise, drank alcohol only in moderation, and never touched cigarettes.

Look at the facts

Smoking kills   About 50 000 people die every year in Britain as a direct result of smoking. This is seven times as many as die in road accidents. Among an average of 1000 young smokers it is likely that one will be murdered, six will be killed in road accidents, and 250 (one-quarter) will die before the end of their natural life-span from diseases caused by smoking.

Smoking causes diseases   90 per cent of lung cancers are caused by smoking. If you smoke five cigarettes a day you are about six times more likely to die of lung cancer than a non-smoker, and if you smoke 20 a day the risk is 19 times greater. Giving up smoking reduces the risk (Fig. 1). Tobacco smoke also causes cancer of the mouth, throat, and wind-pipe. 95 per cent of people who suffer from bronchitis are smokers. Smokers are between two and three times more likely to die of heart disease than non-smokers. Nicotine in smoke increases the heart-beat but narrows blood vessels. Carbon monoxide in smoke reduces the amount of oxygen blood can carry, and may cause fat to be deposited in artery walls. Cigarette smoke kills certain white blood cells. These dead cells release digestive juices which dissolve lung tissue causing thin patches and holes. This disease is called emphysema.

Smokers smell   The breath, clothes, and homes of smokers have an offensive smell. Smokers do not realize this because smoking reduces their sense of smell. It also reduces their sense of taste.

Smoking and pregnancy   Babies of women who smoke during pregnancy are, on average, 200 g lighter than babies of non-smokers. But if a woman gives up smoking while she is pregnant she is likely to have a baby of normal weight. The risk of a baby dying before birth or shortly after is increased by 35 per cent if the mother smokes 20 or more cigarettes a day during pregnancy. Smoking also increases the risk of bleeding during pregnancy, and complications during birth. It may also reduce the production of breast milk.

Smokers harm other people   Children of smokers are more likely to develop bronchitis and pneumonia. In one hour in a smoky room non-smokers inhale as much cancer-causing substance as if they had smoked 15 filter-tipped cigarettes. This small amount of cancer-causing substance can increase the rate of cancer in non-smokers by one-third. The eyes, nose, and throat of a non-smoker are irritated by chemicals in tobacco smoke.

The cost of smoking   The Government collected £3350 million in tobacco taxes in 1980–1. But smoking also *costs* the country money. The cost of medical treatment for people with smoking related diseases amounts to £2 million a week. Added to this is the cost of paying sickness benefit, and widow's pensions and family social security benefits to the dependants of those who die as a result of smoking. There is also the cost to industry and business of the 50 million working days lost each year because of poor health caused by smoking. Furthermore, 20 per cent of all industrial fires are caused by discarded cigarette ends etc., and it has been found that smokers have a higher rate of industrial accidents than non-smokers. 40 per cent of smokers die before retiring age compared with 15 per cent of non-smokers.

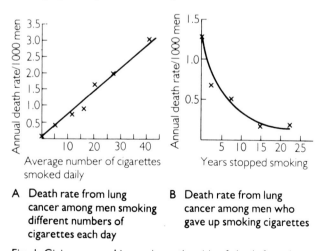

A   Death rate from lung cancer among men smoking different numbers of cigarettes each day

B   Death rate from lung cancer among men who gave up smoking cigarettes

Fig. 1 Giving up smoking reduces the risk of death from lung cancer.

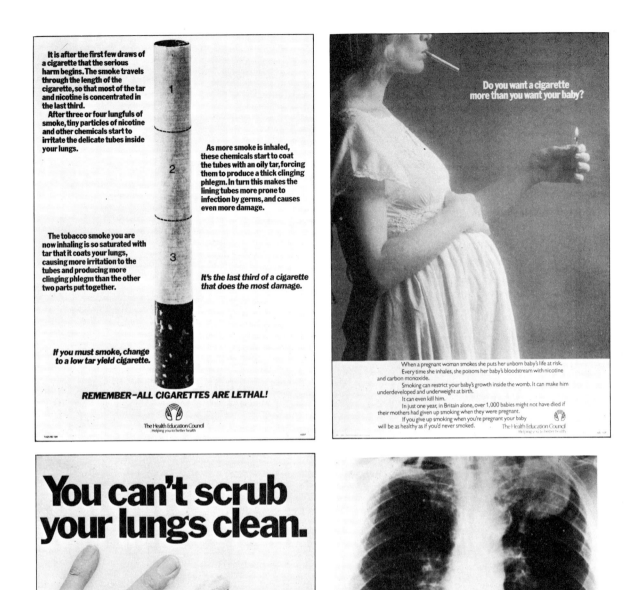

It is after the first few draws of a cigarette that the serious harm begins. The smoke travels through the length of the cigarette, so that most of the tar and nicotine is concentrated in the last third.

After three or four lungfuls of smoke, tiny particles of nicotine and other chemicals start to irritate the delicate tubes inside your lungs.

As more smoke is inhaled, these chemicals start to coat the tubes with an oily tar, forcing them to produce a thick clinging phlegm. In turn this makes the lining tubes more prone to infection by germs, and causes even more damage.

The tobacco smoke you are now inhaling is so saturated with tar that it coats your lungs, causing more irritation to the tubes and producing more clinging phlegm than the other two parts put together.

It's the last third of a cigarette that does the most damage.

If you must smoke, change to a low tar yield cigarette.

REMEMBER—ALL CIGARETTES ARE LETHAL!

The Health Education Council
Helping you to better health

Do you want a cigarette more than you want your baby?

When a pregnant woman smokes she puts her unborn baby's life at risk.

Every time she inhales, she poisons her baby's bloodstream with nicotine and carbon monoxide.

Smoking can restrict your baby's growth inside the womb. It can make him underdeveloped and underweight at birth.

It can even kill him.

In just one year, in Britain alone, over 1,000 babies might not have died if their mothers had given up smoking when they were pregnant.

If you give up smoking when you're pregnant your baby will be as healthy as if you'd never smoked.

The Health Education Council
Helping you to better health

You can't scrub your lungs clean.

Lung cancer kills ten times more smokers than non-smokers.

The Health Education Council Helping you to better health

The white area at the top right corner is a growth on this man's left lung. The lungs are also larger than normal. Compare these unhealthy lungs with the healthy ones on page 71.

9.4 Looking after your body

Look after your body, it's the only one you have.

Body maintenance

Study Figure 1 and list all the ways in which the people are maintaining a healthy body.

Body neglect

Study Figure 2, then list all the ways in which the people are neglecting and ill-treating their bodies. In what way could each item on your list harm the body?

Fig. 1 Body maintenance

Fig. 2 Body neglect

9.5 New bodies for old

Figure 1 describes some of the changes which occur as people get older. Why do these changes occur? Can they be slowed down? How can bionics (bio-engineering) help fight old age?

Diet and ageing

If rats and other laboratory animals have their food reduced to just above starvation level, or if they are made to fast regularly, their lives can be extended by 20 per cent. In humans this would mean another ten to fifteen years of life.

It may be that excess food, or the wrong types of food, eventually clog up body chemistry causing damage which results in ageing. Does this mean over-eaters are 'digging their graves with their teeth'?

Anti-age pills

The body can detect and destroy foreign substances, like germs, which enter it. There is evidence, however, that ageing may occur because this vital self-defence system eventually begins ignoring germs and starts destroying body cells.

Chemicals have now been discovered which appear to stop this act of self-destruction. One day it may be possible to make from these chemicals anti-ageing pills which will allow us to retain youthful health into advanced old age.

Bionics

Bionics, or bio-engineering, is the manufacture of mechanical and other devices which assist or replace damaged or worn-out parts of the body. Here are a few examples which have been, or soon will be, invented.

Artificial skin   Severe burns can now be protected with a chemical solution which is sprayed on the damaged area and sprinkled with a fine powder. The result is a soft, artificial skin which stretches, breathes, and keeps out germs.

Artificial joints   Joints made of stainless steel and plastic are now available to replace leg and arm joints (Fig. 3). Special cement is used to attach them to the bones.

Bionic arms and legs   Scientists are now perfecting bionic limbs which pick up electrical impulses in the stump of a severed limb. These impulses contain all the information needed to move a real limb. A tiny computer in a bionic arm, for example, will use these impulses to control motors which twist the wrist, move the fingers, etc. (Fig. 4). In fact a bionic arm's owner will only have to think of a movement to perform it. Pressure-sensitive electronic skin is now being developed so that bionic limbs can feel things.

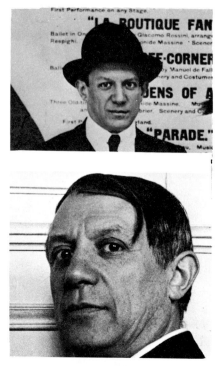

Fig. 1 As humans age: the skin wrinkles; muscles lose weight and power; bones become brittle; the heart pumps less blood; taste, sight, and hearing decline; lungs pass less air; and the brain loses millions of nerve cells. These photos of Picasso show how his face aged throughout his life.

Electronic eyes Micro-electronics will one day be used to make television cameras the size of hazel nuts. These could be fitted into the eye socket in place of damaged or worn-out eyes. The cameras will be connected to a micro computer in special spectacles which will send impulses to electrodes implanted in the visual area of the brain. Here, the impulses will be transformed by the brain into vision.

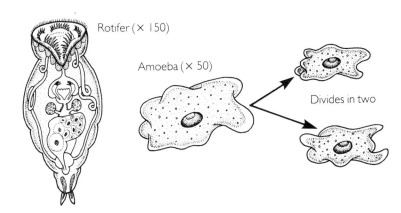

Rotifer (× 150)

Amoeba (× 50)

Divides in two

Fig. 2 Immortal amoebas? Rotifers live for only a few days, dying of 'old age' within hours of laying their eggs. But before an *amoeba* shows signs of old age it divides into two, then these daughter *amoebas* do the same, and so on – forever! Is this a type of immortality?

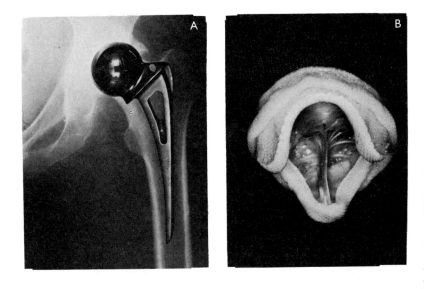

Fig. 3 Spare part surgery A: Many thousands of diseased hip joints have been replaced with steel and plastic joints. The head of the diseased thigh bone is replaced with a stainless steel ball. This fits into a plastic cup attached to the hip bone. **B:** An artificial heart valve can be inserted in the heart to control blood flow.

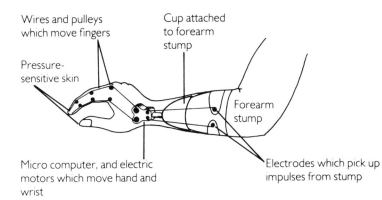

Wires and pulleys which move fingers

Cup attached to forearm stump

Pressure-sensitive skin

Forearm stump

Micro computer, and electric motors which move hand and wrist

Electrodes which pick up impulses from stump

Fig. 4 Bionic limbs of the future will be controlled by micro-computers which pick up electrical impulses from the severed stump. Pressure-sensitive electronic skin will allow the limb to feel objects it touches.

Topic 9 exercises

Health in general

1 Sort the following statements into those which describe someone who is trying to encourage healthy living in others, and someone who is not thinking about the health of others:

a) Not only smokes but offers cigarettes to others.

b) Gives parties at which both non-alcoholic and alcoholic drinks are available.

c) A driver who asks passengers to wear a seat belt.

d) Frequently give sweets to their own as well as other people's children.

e) Stay away from school, work, parties, etc., if they have a bad cold or flu.

f) Smoke without asking permission in other people's homes.

g) Sympathize with people who say they cannot lose weight or give up smoking.

h) Compliment people who are trying to give up smoking and lose weight.

i) Never wash their hands before serving food to others.

2 Explain what is 'good' or 'bad' about each of the people described by the statements in exercise 1.

3 Why is it unhealthy to:

a) use another person's comb?

b) use a bus when you have the time to walk?

c) skip breakfast and eat snacks all morning?

d) wear the same pair of socks for more than a day?

e) eat lots of fatty, sugary, and starchy foods?

f) eat white bread rather than wholemeal bread?

g) read in poor light?

h) spend all day in a stuffy, badly ventilated room?

i) sleep in a damp bed?

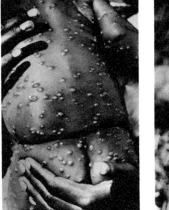 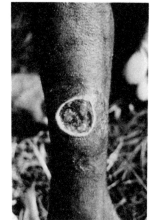

Left: This baby is infected with smallpox
Right: A septic wound

Smoking and health

4 Calculate the cost of smoking 20 cigarettes a day for one year.

5 List five diseases which occur more often in smokers than non-smokers.

6 What effect does cigarette smoke have on white blood cells and the heart blood vessels?

7 List the ways in which smokers make themselves unpleasant and even unhealthy to be with.

8 Why should women never smoke during pregnancy?

9 Why is it never too late to give up smoking?

Giving up smoking

The Health Education Council offers the following advice to those who wish to give up smoking. Try to explain the reasoning behind each piece of advice.

1 First, make up your mind that you really mean to stop smoking. A half-hearted attempt is doomed to failure. (If you can't make up your mind read Unit 9.3. This should help.)

2 Most people find it best to stop smoking suddenly and completely rather than to cut down gradually.

3 After smoking your last cigarette make sure you have none left. Destroy the rest, preferably in front of your family or friends.

4 Tell all your family and friends you have given up smoking, and ask them to help by not offering you any in the future.

5 When you feel the need for a cigarette:

a) Try eating fresh fruit, nuts, crisps, raw carrot, even low calorie sweets.

b) Go for a walk.

c) Try deep breathing to relieve tension.

6 Find friends who also want to stop smoking and persuade them to join you.

7 Vitamin C helps to rid the body of nicotine so drink plenty of fresh orange juice.

8 Develop new hobbies and interests which will take your mind off smoking.

9 Put all the money you would have spent on cigarettes in a money box and count it from time to time.

10 Remember, within days of giving up smoking your breath will lose its offensive smell, your senses of taste and smell will improve, and you will be less short of breath after exercise.

Index and glossary

The combined index and glossary gives brief definitions of some technical words used in the text, followed by the page numbers to which you should turn to see the words in use. Page numbers in heavy type refer to the most detailed explanation. Words in *italic* within a definition are themselves defined elsewhere in the index–glossary.